THE REFRIGERATION & AIR CONDITIONING BIBLE

How to Master the Fundamentals of HVAC Systems and Save Money on Repairs and Maintenance in the Long Run
Includes Valuable Tips for Reducing Costs Right Away

By Alex M. Aguilar

© **Copyright Alex M. Aguilar 2024 - All rights reserved.**

The content contained within this book may not be reproduced, duplicated, or transmitted without direct written permission from the author or the publisher.

Under no circumstances will any blame or legal responsibility be held against the publisher, or author, for any damages, reparation, or monetary loss due to the information contained within this book. Either directly or indirectly.

Legal Notice:

This book is copyright protected. This book is only for personal use. You cannot amend, distribute, sell, use, quote, or paraphrase any part, or the content within this book, without the consent of the author or publisher.

Disclaimer Notice:

Please note the information contained within this document is for educational and entertainment purposes only. All effort has been executed to present accurate, up-to-date, and reliable, complete information. No warranties of any kind are declared or implied. Readers acknowledge that the author is not engaging in the rendering of legal, financial, medical, or professional advice. The content within this book has been derived from various sources. Please consult a licensed professional before attempting any techniques outlined in this book.

By reading this document, the reader agrees that under no circumstances is the author responsible for any losses, direct or indirect, which are incurred as a result of the use of the information contained within this document, including, but not limited to, — errors, omissions, or inaccuracies.

CONTENTS

Part I: Fundamentals of HVAC System	19
Chapter 1: Fundamentals of Refrigeration and Air Conditioning	20
Understanding the Principles of Heat Transfer	**21**
Conduction Heat Transfer	21
Convective Heat Transfer	22
Radiative Heat Transfer	22
Combining Conduction, Convection, and Radiation	23
Heat Exchangers	23
Change of State and Latent Heat	23
Thermal Equilibrium	24
R-value and Thermal Resistance	24
U-factor and Thermal Transmittance	24
Effects of Moisture, Humidity, and Mass	25
Refrigeration Cycle Demystified	**25**
Vapor-Compression Cycle Overview	26
Compressor	26
Condenser	27
Metering Device	28
Evaporator Coils	28
Refrigerants	28
Refrigeration Cycle Analysis	29
Refrigeration Load Matching	29
Coefficient of Performance	29
Vapor Compression Variations	30

Types of HVAC Systems: Exploring Variations — 30
- Split vs Packaged Systems — 31
- Central HVAC Systems — 31
- Mini-Split Systems — 32
- Variable Refrigerant Flow (VRF) — 32
- Geothermal Heat Pumps — 32
- Chilled Beams — 33
- Thermal Storage Systems — 33
- Radiant Heating and Cooling — 33
- Ventilation & Air Filtration Systems — 34
- Customized & Integrated Systems — 34

Key Components: In-Depth Analysis — 34
- Compressors — 34
- Heat Exchangers — 35
- Fans and Blowers — 35
- Pumps — 36
- Condensers and Evaporators — 36
- Thermostatic Expansion Valves — 37
- Refrigerant Piping — 38
- Electrical Controls — 38

Navigating the Jargon: A Comprehensive Glossary — 38
- Thermodynamics Terms — 39
- Refrigerant Terms — 40
- HVAC Equipment Terms — 41
- HVAC Configuration Terms — 42
- HVAC Performance Terms — 42
- HVAC Distribution Terms — 43
- HVAC Operations Terms — 44
- HVAC Load and Performance Calculations — 44
- HVAC Mechanical Equipment — 45
- HVAC Safety Essentials — 46
- HVAC Maintenance Essentials — 46
- HVAC Commissioning Concepts — 47
- Electrical and Controls — 47
- HVAC Design and Analysis — 48
- HVAC Heating Equipment — 49
- HVAC Cooling Equipment — 49
- HVAC Ventilation Essentials — 50
- HVAC Environments and Air Quality — 51
- Building Automation Systems — 51

Chapter 2: Installation and Maintenance Guidelines — 53
Step-by-Step HVAC Installation Guide — 53
Assessing Your Space: Sizing and Load Calculation — 54
Selecting the Right HVAC System: Factors to Consider — 55
Ductwork Design and Installation: Ensuring Efficiency — 58
Electrical and Plumbing Requirements: A Comprehensive Overview — 60
Safety Precautions During Installation: Protecting Your Investment — 61
Commissioning and Testing: Verifying System Performance — 63

Comprehensive Routine Maintenance for Peak Performance — 65
Monthly Filter Changes: Choosing the Right Filters — 65
Cleaning and Lubricating Moving Parts: Fan Blades, Bearings, and More — 67
Coil Cleaning: Evaporator and Condenser Coils — 69
Refrigerant Level Checks: Avoiding Overheating or Freezing — 71
Inspection of Electrical Components: Wiring, Capacitors, and Contacts — 72
Calibrating Thermostats: Accuracy Matters — 74

Seasonal Maintenance Checklist — 76
Spring Startup: Preparing for Cooling Season — 78
Fall Shutdown: Getting Ready for Heating Season — 79

Troubleshooting Common HVAC Problems — 80
Inadequate Cooling or Heating: Potential Causes and Solutions — 80
Uneven Temperature Distribution: Balancing Airflow — 81
Strange Noises: Identifying and Addressing Unusual Sounds — 82
Leaking Refrigerant: Detection and Resolution — 83
Thermostat Malfunctions: Diagnosis and Repair — 84
Electrical Issues: Safety Measures and Solutions — 85

DIY vs. Professional Maintenance — 86
Knowing Your Limits: When to Call in a Professional — 86
Finding a Reliable HVAC Technician: Hiring Tips — 87
Cost-Benefit Analysis: DIY vs. Professional Maintenance — 89
Developing a Long-Term Maintenance Plan: Balancing DIY and Professional Services — 90
Preventive Maintenance Contracts: Pros and Cons — 91

Part II: Efficiency and Savings — 93

Chapter 3: Energy Efficiency and Cost Savings — 94
Mastering Energy Efficiency: Strategies and Tactics — 95
Educating Homeowners — 95
Thermostat Management — 95
Regular System Check-Ups — 95
Filter Maintenance — 96
Duct Assessment and Sealing — 96

HVAC Sizing Fundamentals	96
Zoning System Basics	96
Detecting Duct Leaks	96
Balancing the Distribution System	97
Conducting Load Calculations	97
Creating a Home Energy Profile	97
Using Energy Monitoring Tools	97
Infrared Imaging Basics	97
Upgrading Your System for Long-Term Savings	**98**
Replacing Aging Equipment	98
High-Efficiency Equipment Options	98
Gas vs. Electric Equipment Efficiency	98
Air Conditioner Efficiency Upgrades	98
Furnace Fan Upgrades	99
Proper Equipment Sizing	99
Zoned Heating Systems	99
Ductless Mini-Split Heat Pumps	99
Geothermal Heat Pump Systems	99
Heat Recovery Ventilation Systems	100
Duct Insulation and Sealing	100
Variable Capacity HVAC Options	100
Heat Pump Water Heaters	100
HVAC Control Upgrades	100
Building Envelope Improvements	100
HVAC Replacement Prioritization	101
HVAC Replacement Project Planning	101
Smart Controls: The Power of Programmable Thermostats	**101**
Evaluating Current Controls	101
Programmable vs. Smart Thermostats	102
Programming Fundamentals	102
Achieving Optimal Comfort	103
Customizing Setback Strategies	103
Geo-Locating Smart Thermostats	103
Programming for Vacations	104
Optimizing Thermostat Placement	104
Maximizing Energy Savings Features	104
Home Insulation and Sealing Techniques	**105**
Attic Insulation	105
Wall Insulation	106
Floor Insulation	106
Potential Energy Savings from Sealing and Insulating	106

Identifying Air Leaks	106
Sealing Methods	107
Duct Sealing Techniques	107
Balancing the Distribution System	107

Financial Benefits: Tax Credits and Incentives — 107
- HVAC Equipment Rebates — 107
- Smart Thermostat Rebates — 108
- Home Performance Rebates — 108
- Heat Pump Rebates — 109
- Income-Qualified Programs — 109

Chapter 4: Common HVAC Problems and Solutions — 110

A Comprehensive Troubleshooting Toolkit — 111
- Understanding the HVAC System Diagram: Key Components — 111
- Reading Error Codes: Deciphering Diagnostic Messages — 113
- Measuring Airflow: Tools and Techniques — 113
- Thermographic Imaging: Detecting Heat Loss and Cold Spots — 114
- Manometer Usage: Checking Gas Pressure — 115

DIY Fixes for Minor Issues: Step-by-Step — 116
- Clogged Air Filters: Replacement and Cleaning — 116
- Frozen Evaporator Coils: Causes and Solutions — 118
- Condensate Drain Blockage: Clearing the Line — 118
- Pilot Light Problems: Relighting and Safety Measures — 119
- Non-Functioning Thermostat: Calibration and Battery Replacement — 119
- Air Duct Leaks: Identifying and Sealing — 120

When to Call a Professional - Warning Signs — 120
- The Importance of Timely Intervention — 121
- Gas Odors and Leaks: Immediate Action Required — 121
- Electrical Issues: Flickering Lights and Circuit Breaker Trips — 121
- Refrigerant Leaks: Environmental and Efficiency Concerns — 122
- Persistent Strange Noises: Bearing Failure or Fan Issues — 122
- Irreparable Heat Exchanger Damage: Safety Hazards — 122

Preventing Future Problems - Proactive Measures — 123
- The Value of Regular Inspections — 123
- Implementing a Preventive Maintenance Schedule — 123
- Keeping the Outdoor Unit Clean: Debris Removal — 124
- Upgrading Insulation: Energy Efficiency and Comfort — 124
- Installing Carbon Monoxide Detectors: Safety Assurance — 125
- Educating Family Members: Safety and Awareness — 125

Prolonging the Lifespan of Your HVAC System — 126
- Systematic Maintenance: The Key to Longevity — 126

Educating Homeowners: Proper HVAC Usage — 127
Tracking Performance Metrics: Monitoring Efficiency — 127
Considering Upgrades: Energy-Efficient Components — 128
Recycling and Disposal: Environmentally Responsible Practices — 128
Planning for System Replacement: When It's Time — 129

Part III: Comfort and Advanced Techniques — 130

Chapter 5: Indoor Air Quality and Comfort — 131

Prioritizing Indoor Air Quality: Why It Matters — 132
- Health Effects of Poor Indoor Air Quality — 132
- Productivity and Performance Impacts — 133
- Sources of Indoor Air Pollution — 133
- Populations Vulnerable to Indoor Air Pollution — 134
- Benefits of Enhanced Indoor Air Quality — 134

Air Filtration and Purification Technologies — 135
- Filter Fundamentals — 135
- Ionizing and Ozone Air Purifiers — 135
- Activated Carbon Filtration — 136
- Ultraviolet Germicidal Irradiation — 136
- Photocatalytic Oxidation Purifiers — 136
- Whole-Building vs Portable Purifiers — 137

Achieving Ideal Humidity Levels for Comfort — 137
- Impacts of Improper Humidity Levels — 137
- Recommended Humidity Setpoints by Season — 137
- Devices for Measuring Humidity — 138
- Managing Humidity Through HVAC Equipment — 138
- Supplemental Humidity Control Strategies — 138

Zoning Your HVAC System: Room-by-Room Control — 139
- Benefits of Zoned HVAC Control — 139
- Designing a Zoning Layout — 139
- Dampers for Low-Cost Zoning — 140
- Zone Control Boards and Dampers — 140
- Multi-Stage and Variable Equipment — 140
- Mini-Split Zoning Systems — 141

Striking a Balance: Comfort vs. Efficiency — 141
- Design Factors for Balancing Comfort and Efficiency — 141
- Advanced Equipment for Enhanced Control — 142
- Control Strategies to Harmonize Comfort and Efficiency — 142
- Operational Techniques for Balancing Comfort and Efficiency — 142
- Occupant Engagement for Achieving Goals — 143

Commissioning for Quality Assurance 143

Chapter 6: Advanced HVAC Techniques and Technologies 144
Harnessing Earth's Energy: Geothermal HVAC Systems 145
Types of Geothermal Heat Pumps 146
Geothermal Heat Pump Operation 147
Real-World Applications and Examples 149
Installation, Maintenance, and Cost Considerations 151
Sustainability and Savings: Solar-Powered HVAC Systems 152
Types of Solar Thermal Collectors 152
Solar HVAC System Types 153
Solar HVAC Applications and Case Studies 154
Efficiency Redefined: Variable Refrigerant Flow (VRF) HVAC Systems 155
Key Components and Operation 155
Applications and Usage Examples 156
Benefits, Savings Potential, and Challenges 157
Fresh Air Solutions: Energy Recovery Ventilation (ERV) 158
How ERVs Work 158
Components and Operation 159
ERV Efficiency & Benefits 160
ERV Applications & Sizing 161
Application Case Study 162
Future Trends in HVAC: What Lies Ahead 163
Artificial Intelligence and Machine Learning 163
Connected Buildings 164
Renewable Integration 165
Grid Harmonization 165
Carbon-Free Future 166

Part IV: Cost Reduction & Real-Life Examples 167

Chapter 7: Tips for Immediate Cost Reduction 168
Quick Wins for Immediate Energy Bill Reduction 169
Replace Air Filters Regularly 169
Clean or Replace the Furnace Vent 169
Check Ductwork for Leaks 170
Insulate Exposed Ducts and Pipes 171
Adjust Thermostat Settings 171
Lower Window Treatments 172
Lubricate Fan Motors 173
DIY Projects for Instant Savings 173

Clean or Replace Furnace Burners ... 173
Change Thermostat Batteries ... 174
Clean Condensate Drains ... 175
Inspect Heat Exchanger ... 175
Negotiating with HVAC Contractors: Insider Tips ... **176**
Research Average Local Rates ... 176
Request Service Plan Details ... 176
Ask for Referral Contact Information ... 176
Inquire About Financing Options ... 177
Bundle Maintenance with Installations ... 177
Get Agreement Details in Writing ... 177
Understanding Your Energy Usage: Tools and Resources ... **177**
Review Past Energy Bills ... 177
Analyze Online Energy Portal Data ... 178
Schedule a Home Energy Audit ... 178
Review Smart Thermostat App Data ... 178
Benchmark Against Similar Homes ... 178
Track Individual Circuit Usage ... 179
Creating a Customized HVAC Maintenance Schedule ... **179**
Spring Startup (Late March-May) ... 179
May (Annual) ... 179
Summer (June-August) ... 180
Fall (September-October) ... 180
Winter (November-February) ... 180

Chapter 8: Case Studies and Success Stories ... 181
Case Study: Annual Savings of $2,000 Through Preventative Maintenance ... 182
Scheduling Regular Checkups ... 182
Replacing Aging Components ... 182
Improving Airflow and Filtration ... 182
Optimizing Settings and Controls ... 182
Measuring Energy Savings ... 183
Renewing the Maintenance Plan ... 183
Spreading the Word ... 183
Valuable Lesson Learned ... 183
Referrals Yield More Business ... 183
Preventative Maintenance Pays Dividends ... 183
Expert Insight: An HVAC Contractor's Time-Tested Strategies ... 184
Overhaul Saves Thousands for a Small Business ... 185
Home Efficiency Upgrades: A Family's Dry Climate Success ... 186
Renovating a Historic Home's HVAC Suite ... 187

Assessing Aging Equipment	187
Plumbing Modern Upgrades	187
Selecting an Efficient Heating System	187
Preserving Historic Character	188
Construction and Installation	188
Unparalleled Comfort Gains	188
Delighted with Efficiency	188

Part V: Advanced HVAC Insights — 189

Chapter 9: Commercial HVAC Systems — 190

Commercial HVAC Basics: Key Differences — 191
- Size and Scale — 191
- Multiple HVAC Zones — 192
- Higher Performance Requirements — 192
- Year-Round Usage — 193
- More Complex Control Systems — 194
- Importance of Air Quality — 194

Sizing and Design Considerations — 195
- Load Calculations — 195
- Equipment Selection — 195
- Air Distribution Design — 196
- Controls Layout — 196
- System Placement — 197
- Commissioning and Testing — 197

Energy Efficiency in Commercial Settings — 198
- High Efficiency Equipment — 198
- Active Controls — 198
- Heat Recovery Systems — 199
- Economizers — 199
- Insulation and Air Sealing — 200
- Lighting/Daylight Harvesting — 200
- Renewable Energy Integration — 200

Maintenance Challenges and Solutions — 201
- Continuous Operation — 201
- Complex Component Access — 201
- Specialized Skills — 202
- Critical Nature of Operations — 202
- Tight Indoor Environments — 202
- Inter dependencies — 203
- Changing Codes & Standards — 203

Preventative Maintenance	203
Training & Safety	203
Case Studies in Commercial HVAC Success	**204**
Office Renovation Boosts IAQ and Cuts Costs	204
Green Retrofit for Retail Chain	204
Delivering Comfort via Geothermal	205
Retro-commissioning Pays Off	205
University Campus Goes Green	205

Chapter 10: HVAC Regulations and Environmental Impact — 206

Regulatory Landscape: Understanding Compliance — **207**
- Regulatory Bodies — 207
- Installation Standards — 207
- Permitting Requirements — 208
- Record-Keeping Requirements — 208
- Ongoing Training Requirements — 209
- Energy Efficiency Regulations — 209
- Special HVAC Situations Requiring Regulatory Consideration — 210

Environmental Impact of HVAC Systems — **211**
- Refrigerant Management — 211
- Emission Regulations — 212
- Energy Efficiency Standards — 213
- Building Energy Codes — 213
- Environmental Permitting — 214

Green HVAC Technologies: Sustainability Focus — **214**
- High Efficiency Equipment — 215
- Renewable Energy Integration — 215
- Demand Response Enrollment — 216
- Zero Emission HVAC Options — 216
- Advanced System Optimization — 217

Future Regulations and Industry Trends — **217**
- Refrigerant Phase-Downs — 217
- Net-Zero Carbon Buildings — 218
- Climate Change Mitigation Policies — 218
- Decarbonization of Electricity Grids — 218
- Increasing Resiliency to Climate Disruption — 219
- Decarbonized Energy for Transportation — 219
- Digitalization Transforming Building Operations — 220

Navigating Compliance for Cost Savings — **220**
- Adopting Efficient, Eco-Friendly Technologies — 220
- Taking Advantage of Tax Incentives — 220

Strict Documentation Practices 221
Ongoing Regulatory Education 221
Preventative Maintenance Programs 221
Special Compliance Considerations 222
Pursuing Sustainability Accreditation 222

Conclusion 224

Appendices 227

Glossary of HVAC Terms **227**
Recommended Tools and Equipment **229**
Worksheets and Checklists **231**
   HVAC Installation Checklist 231
   Monthly HVAC Maintenance Checklist 232
   HVAC Troubleshooting Checklist 232
   HVAC Energy Efficiency Checklist 233

INTRODUCTION

Welcome to The Refrigeration & Air Conditioning Bible - your definitive guide to mastering HVAC systems. With over 20 years in the HVAC industry as a technician, business owner, and educator, I've seen firsthand how HVAC mastery leads to lower costs, improved efficiency, and greater peace of mind. This book contains the invaluable insights and practical skills I've developed throughout my career installing, troubleshooting and maintaining all types of HVAC systems in homes and businesses across the country.

My journey into the world of climate control systems began over two decades ago when I got a summer job with my uncle's HVAC company in Phoenix, Arizona. As a curious teenager, I peppered the experienced technicians with endless questions about the intriguing pipes, wires, valves and metal boxes I'd seen them hauling to and from job sites. The complexity and precision required to create an invisible blanket of perfectly tempered air soon had me hooked! Before long, I traded my fast food gig for an HVAC apprenticeship and never looked back.

Over the next four years, I logged thousands of hours installing, servicing and repairing air conditioning, heating and refrigeration units of all makes and models. Each job was an invaluable real-world classroom for decoding the science behind HVAC and developing the troubleshooting skills every technician needs in their toolkit. Whether figuring out why an AC unit iced over or a furnace wouldn't fire, I treated every service call like a mystery to untangle. Solving customers' temperature troubles quickly became a passion.

To deepen my arsenal of problem-solving strategies and master the newest technologies hitting the market, I crisscrossed the country attending advanced HVAC training seminars after earning my technician certifications. From heat pumps in Houston to boilers in Boston - I set out to understand it all, driven by the satisfaction I took in ensuring everyone's comfort. By experimenting with leading-edge troubleshooting gadgets and quizzing field pioneers on their tricks of the trade, I uncovered time-saving diagnostic shortcuts and perfected my installation methods for optimal efficiency and longevity.

I spent a decade crisscrossing North America on service calls helping homeowners and business owners find the ideal climate control solution for their needs, budget and building. Visiting so many sites gave me invaluable insights into how flaws in ductwork design, insulation or equipment capacity can undermine system functioning that the best HVAC setup combines art and science to account for all the variables.

Eager to equip other aspiring HVAC pros with the complete set of skills to excel, I began teaching others. Teaching the intricate details of realizing optimal temperature, humidity and ventilation in real-world spaces forced me to simplify complex concepts for students from all backgrounds. Over years of refining my teaching approach, I discovered how to breakdown technical elements into accessible building blocks that accelerate learning. Equipping new generations of climate control experts with a complete mastery became my new passion.

Distilling two decades of hands-on HVAC experience into this comprehensive tome has been a labor of love for me. Consider it your playbook for achieving a perfectly calibrated indoor environment via easy-to-deploy knowledge I wish I had on day one. By mastering HVAC fundamentals, developing diagnostic skills, and applying insider efficiency upgrades and cost-reduction tips, you can save significant money while controlling comfort with precision no matter the season or system.

This Refrigeration & Air Conditioning Bible is divided into key sections designed to build your climate control expertise step-by-step:

Part I covers everything you need to know about the inner workings of heating, ventilation and AC units from compressor types to electrics to the pros and cons of various systems on the market today. With key terminology demystified and components explored in detail, you'll have a complete working knowledge to draw upon for all applications.

Part II leverages your new system mastery to boost efficiency and savings via equipment selection/sizing, DIY maintenance, thermostat set points, insulation factors and energy tax incentives. Both immediate cost savings tips and long terms upgrades are covered to decrease your power bills.

Part III unveils the advanced diagnostics and technologies shaping HVAC today, from zoning to geothermal installations. You'll learn how to pinpoint comfort issues, balance air distribution, purify indoor air quality and realize the potential for smart climate control to impact health, productivity and even sleep!

Part IV and V features real-world case studies and success stories from my own experience and other leading HVAC experts that reveal how leveraging the strategies in this book transforms climate control capabilities in homes and businesses. Countless before and after scenarios showcase HVAC mastery in action across industries and locations with dramatic results!

While being an HVAC Pro certainly keeps me busy year-round, when I finally step away from air handlers and heat pumps, you can find me hiking through the mountain trails near my home in Durango, Colorado for some welcome solitude and fresh perspective. The wonders of the natural world never cease to amaze me, whether it's encountering a moose or watching an electrical storm far off over the mesa. The interplay of atmospherics and energy serves as the perfect metaphor for HVAC systems to me.

My hope is that this bible provides you with an equivalent sense of wonder and excitement upon unlocking the inner workings of indoor climate control technology and seeing what it takes to calibrate systems to perfection. By committing yourself to HVAC mastery today, you put yourself on the path towards far greater savings, efficiency and control tomorrow. So dive in and let the journey begin!

PART 1
FUNDAMENTALS OF HVAC SYSTEM
PART 1

CHAPTER 1
Fundamentals of Refrigeration and Air Conditioning

Refrigeration and air conditioning systems have become an integral part of modern life. From keeping our homes comfortable to preserving food and medicine, these technologies allow us to overcome nature's tendencies and create the environment we want. To use refrigeration and AC systems effectively, we must first understand the core scientific principles that enable them to function.

UNDERSTANDING THE PRINCIPLES OF HEAT TRANSFER

Refrigeration and air conditioning technologies rely on leveraging natural tendencies in the movement of heat to achieve cooling. Utilizing heat transfer physics, these systems can collect heat from one area where it is unwanted and successfully move it somewhere else. The three main methods of heat transfer are conduction, convection, and radiation. Mastering the science behind these concepts is key to fully grasping how modern refrigeration and air conditioning systems operate.

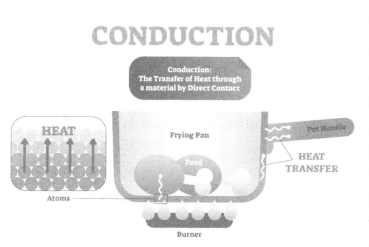

Conduction Heat Transfer

Conduction is transferring heat energy within or between two objects in direct contact. It occurs at the molecular level as faster-moving molecules in a hotter region collide with slower-moving molecules in an adjacent cooler region, transferring some of their kinetic energy. Metals are excellent thermal conductors because their free electrons can quickly transport thermal energy. Insulators like plastic foam limit conductive heat transfer by trapping air pockets and limiting molecular contact. Understanding the thermal conductivity values of different materials is vital in the design of refrigeration system components. Copper and aluminum are often used for heat exchangers because they offer high conductivity to maximize heat transfer. Insulation materials like polyurethane foam are specially chosen to block conductive heat gain. Refrigerant lines must be properly insulated to prevent wasted energy transfer through conduction. Even the conductivity of the metal alloys used in compressors can factor into their overall efficiency and operating temperatures.

The rate of conductive heat transfer follows Fourier's Law, which states that the heat flux through a material is proportional to the temperature gradient and the material's thermal conductivity. This means materials like copper, which have high thermal conductivity, will transfer heat rapidly compared to insulators. The temperature differential, or gradient, is also important - higher gradients drive faster heat transfer.

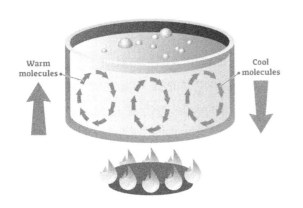

Convective Heat Transfer

Convection is heat transfer by the bulk motion or flow of a fluid like air or water. It relies on the principle that as an object heats up, its molecules move faster and spread apart, becoming less dense. This causes the hotter, less dense matter to rise relative to the cooler, denser matter around it. Hot and cold fluid currents are created as the hot fluid flows up and the cold fluid drops down, circulating and transferring heat from one area to another.

Convection currents play a major role in HVAC systems by enabling air handlers and hydronic heating systems to transfer collected heat from one space to another. Understanding natural and forced convection aids design and troubleshooting. In natural convection, differences in fluid density created by temperature variations drive flow. This allows heat to rise from radiators and hot surfaces. Forced convection uses fans and pumps to circulate fluids actively. Careful sizing and duct or pipe layout promote proper airflow and water flow. Convection transfers heat according to Newton's Law of Cooling, which describes the link between heat flux, fluid temperatures, flow rates, and convective heat transfer coefficients.

Radiative Heat Transfer

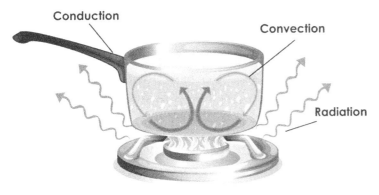

Radiation is the transfer of heat energy by electromagnetic waves directly across space without relying on the bulk flow of matter. All objects above absolute zero continuously emit thermal radiation corresponding to their temperature. The intensity and wavelength of this radiation depend on the object's surface temperature - hotter objects radiate more intensely, especially at shorter wavelengths. When radiation strikes an object, the waves are absorbed by its molecules, exciting increased molecular activity that manifests as heating.

Radiation plays a key role in heat transfer when dealing with heat sources not in direct contact, like the immense thermal radiation emitted by the sun or the heat from a burner flame inside a furnace. Engineers consider radiative heat gains and losses in load calculations when designing HVAC systems. Windows, for example, have a major impact on radiative heat transfer. Low-E coatings selectively reflect infrared rays to reduce gain in summer yet admit visible sunlight in winter.

Combining Conduction, Convection, and Radiation

Practical heat transfer scenarios involve combinations of conduction, convection, and radiation working in tandem: Conduction transfers heat from a furnace combustion chamber to its exterior surfaces. Convection carries heat into air flowing from the furnace plenum to ductwork. Radiation emits some heat into the mechanical room from the hot furnace exterior. Duct conduction and air convection distribute hot air to spaces. Room surfaces like windows absorb radiative solar gains. Occupants warm air via conduction and convection. The warmed air rises through convection, transferring heat to surrounding walls through conduction.

Heat Exchangers

Heat exchangers are vital components in HVAC systems that enable two fluids at different temperatures to exchange heat while keeping them physically isolated. This allows, for example, collected heat from an indoor space to be safely transferred to an outdoor airstream without contamination or mixing. Copper and aluminum are common heat exchanger metals due to their high thermal conductivity, which maximizes heat transfer efficiency between the two fluids. Various exchanger configurations like finned tube, shell and tube, plate and frame, and others are tailored to particular system types and capacities. Careful design minimizes fluid pressure losses while maximizing surface contact area through techniques like small passages or folded fins. Keeping heat exchangers free of fouling buildup maintains performance, so regular inspection and cleaning are key.

Change of State and Latent Heat

Change of state, or phase change, between liquid and gas results in a significant heat transfer known as latent heat. The latent heat of vaporization is the energy required to boil a liquid into vapor. In contrast, the latent heat of condensation is the heat released when a vapor condenses back into a liquid. Water's unusually high latent heat values, over 1000 BTU/lb, make it ideal for heat storage applications. Refrigerants are likewise optimized for how much heat they can absorb or release during phase change. Latent heat transfer underpins condensation and evaporation in refrigeration cycles, so engineers carefully select refrigerants suited for target temperature ranges.

Thermal Equilibrium

Thermal equilibrium is when two objects or systems cease to exchange net heat energy because their temperatures have equalized. Refrigeration aims to sustain a state of disequilibrium by continuously adding energy into or extracting heat from a system faster than equilibration occurs. This offsets the natural tendency towards equilibrium, stabilizing room temperature at surrounding ambient levels. Components like sensors and PID controllers using feedback loops automatically adjust system operation to maintain the desired disequilibrium state and temperature differential.

R-value and Thermal Resistance

R-value and its inverse, C-value, quantify a material's thermal resistance or how well it resists conductive heat transfer. Materials with higher R-Values, like insulation boards, provide greater resistance to heat flow, while more conductive metals and glass demonstrate lower R-Values. Summing R-Values of composite materials estimates total assembly resistance. High-resistance insulators reduce heat transfer, while low-resistance conductors promote it. Technicians consider R-Values when selecting insulation thicknesses needed for various system components. Engineers consult R-value tables when calculating heat loads for sizing equipment. Understanding thermal resistance helps control conductive heat flows to maximize efficiency.

U-factor and Thermal Transmittance

While the R-value measures a material's resistance to conductive heat flow, the U-factor quantifies its inverse property - thermal transmittance - indicating how readily heat passes through. U-factor, also known as U-value, represents the amount of heat in BTUs transferred through one square foot of a construction assembly for one hour per one degree F of the temperature difference across the assembly.

Lower U-factors indicate better-insulated components that allow less heat transfer. Comparing U-factors helps engineers select advantageous windows, walls, roofing materials, and insulation for HVAC efficiency. Technicians apply U-values when identifying causes of unexpected heat gain or loss.

Effects of Moisture, Humidity, and Mass

Moisture and humidity influence heat transfer in HVAC systems and spaces both directly and indirectly:

Latent heat required to evaporate or condense moisture directly absorbs or releases significant energy. Humidity affects material conductivity - concrete is 5X more conductive when wet. More dense materials like water convey heat faster than less dense air. Moisture migrates through permeable materials via wicking.

Understanding moisture's impacts allows proper humidification and dehumidification control strategies and mitigating issues like condensation, corrosion, and mold. Considering air density and specific heat capacity differences evaluates ventilation effectiveness. Moisture behavior knowledge ensures optimal comfort and system reliability.

The refrigeration cycle is the continuous process utilized in refrigerators, air conditioners, and heat pumps to collect heat from one area and reject it in another. This physics-based cycle, comprised of key components undergoing thermodynamic changes, establishes a closed-loop sequence for heat transfer. By thoroughly understanding each part of the refrigeration cycle, we gain insightful comprehension of the inner workings of cooling systems.

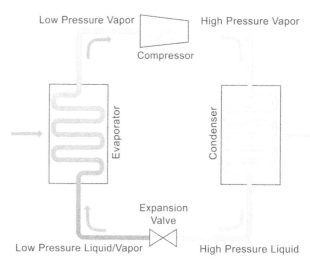

Vapor-Compression Cycle Overview

The vapor-compression cycle is the most common refrigeration cycle configuration. It relies on phase-changing refrigerant to absorb, relocate, and dissipate heat through sequential processes:

1. The low-pressure refrigerant vapor enters a compressor where it is mechanically compressed, raising its temperature and pressure.
2. Hot, high-pressure vapor flows into a condenser where heat is rejected to ambient air or water, condensing the vapor into a liquid.
3. The high-pressure liquid passes through an expansion device, undergoing a sharp pressure and temperature drop.
4. Low-pressure cold liquid re-enters the evaporator coil, absorbing heat from the enclosed space and evaporating back into a vapor, resuming the cycle.

This continuous sequence works to collect undesired heat and discharge it into the environment. Understanding the objectives and functions of each component reveals how cooling is achieved.

Compressor

The compressor is the "heart" of the refrigeration cycle, circulating and pressurizing the refrigerant vapor. Compression raises the refrigerant's temperature to facilitate heat rejection. Common compressor types each have advantages:

- **Positive displacement** - Reciprocating or rotary mechanisms increase pressure through volume reduction.
- **Dynamic** - Impellers accelerate the flow to induce pressure rise through velocity increase.
- **Absorption** - Use thermal compression driven by external heat sources.

Capacity modulation matches loads. Proper sizing prevents inefficient over or under-compression. Keeping compressors well-maintained maximizes reliability and performance over long operating lifetimes.

Condenser

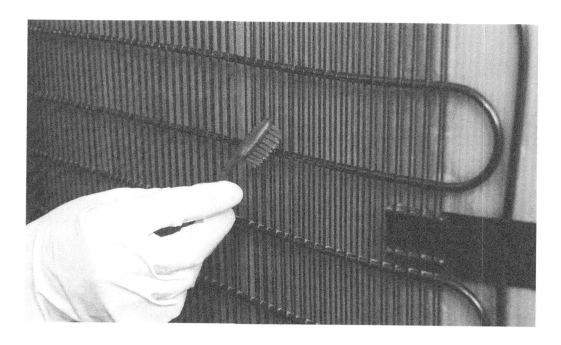

The condenser removes heat from the hot, pressurized refrigerant vapor, causing it to condense into a liquid state. Air or water flows across the condenser coils or tubes, absorbing emitted heat via conduction and convection. Fins provide added surface area to enhance heat transfer to the surroundings. High-efficiency condensers maximize heat rejection while minimizing refrigerant pressure losses. Keeping condensers free of dirt buildup maintains optimal performance. Properly designed, sized, and located condensers are key to cycle efficiency.

Metering Device

Metering devices like capillary tubes or electronic expansion valves regulate refrigerant flow from the high-pressure liquid state at the condenser outlet to the low-pressure liquid state entering the evaporator. This dramatic pressure reduction results in explosive liquid flashing and extreme temperature drop. Controlling this expansion avoids excessive refrigerant flow while ensuring adequate supply. The degree of refrigerant "subcooling" is monitored to check proper operation. Metering devices also help prevent liquid refrigerant from reaching the compressor.

Evaporator Coils

The evaporator coil absorbs heat from the air or liquid being cooled, causing the low-pressure liquid refrigerant flowing through its tubes to boil and evaporate into a vapor. This state change from liquid to gas absorbs substantial latent heat, cooling the refrigeration cycle. Air blown across cooling coils removes humidity while transferring heat from the cooled space or substance. Optimal coil design maximizes surface contact with circulating air while maintaining sufficient airflow. Keeping coils free of frost buildup ensures efficiency.

Refrigerants

Refrigerants are heat transfer fluids vital to the refrigeration cycle. Their properties are tailored to application requirements:

- Volatility facilitates phase change between liquid and vapor states.
- Thermal capacity maximizes heat absorption and dissipation.
- Safety, environmental impacts, efficiency, and cost are also considered.

Common refrigerants include hydrofluorocarbons like R-410a and hydrocarbons like R-290. Utilizing the optimal refrigerant improves system performance. Proper handling is critical, including pressure regulation, containment, and recovery practices.

Refrigeration Cycle Analysis

Analyzing refrigeration cycles involves considering the following:
- Pressure-enthalpy diagrams show the thermodynamic progression of refrigerants through phase change processes. Comparing to expected curves identifies issues.
- Temperatures and pressures at key points checked against specifications and ambient conditions indicate proper operation.
- Subcooling and superheat margins verify adequate refrigerant metering and evaporator performance.
- Comparing theoretical refrigeration effect to actual cooling produced checks for losses.

This technical cycle analysis helps maximize efficiency and troubleshoot problems.

Refrigeration Load Matching

A careful refrigeration system design matches the required cooling capacity to the anticipated load while providing a reasonable safety factor or buffer. Key factors in determining load:
- Volume - Size of space requiring cooling.
- Occupancy - Internal gains from people, equipment, and processes.
- Activities - Cooking and bathing add major loads to homes.
- Outdoor conditions - Temperature, humidity, and peaks.
- Construction materials - Insulation levels and air infiltration influence gains.

Proper load balancing prevents short cycling from oversized systems yet assures comfort with adequate capacity. Right-sizing saves energy.

Coefficient of Performance

The coefficient of performance (COP) measures refrigeration cycle efficiency. It is calculated by dividing cooling output in BTUs by the compressor input work in watts. The higher the COP, the more cooling is obtained per unit of energy. COP depends on factors like compressor and heat exchanger effectiveness, cycle design, and operating conditions. Comparing rated to actual COP checks for efficiency loss. Optimizing COP reduces operating costs.

Vapor Compression Variations

Unique vapor compression configurations suit particular use cases:

- Cascade cycles use multiple refrigerant circuits at different pressure/temperature levels for an expanded operating range.

- Two-stage cycles add an intermediate pressure stage for increased moisture removal and capacity control.

- Absorption cycles use external heat sources like natural gas flames or steam instead of electric compressors.

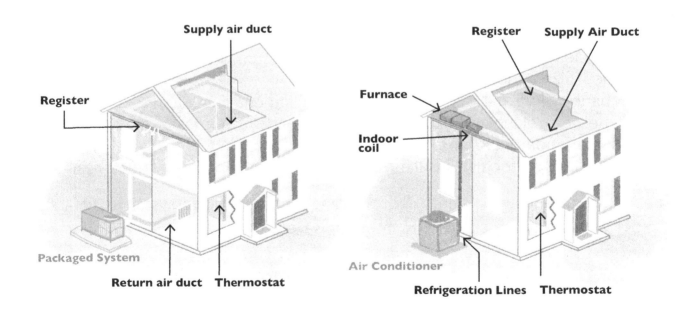

HVAC systems utilize refrigeration cycles tailored for heating and cooling inhabited spaces. Several fundamental design variations exist, each with advantages for certain applications and scenarios. By thoroughly comparing the capabilities, configurations, and components comprising the various HVAC system types, appropriate technology choices can be identified for a given building's needs.

Split vs Packaged Systems

Split and packaged are two main HVAC system configurations: Split systems have separate indoor and outdoor units connected by refrigerant piping and electrical wiring. This allows flexible installation with long piping runs to locate condensers away from occupied spaces.

Split systems can have one or many indoor evaporator units paired with each outdoor condensing unit.

Packaged systems house all components together in one outdoor cabinet. This saves installation time and cost but limits placement to exterior walls with adequate clearances. Packaged units are common for small commercial buildings. Various capacities and configurations are available.

Comparing installation factors, operating costs, and reliability helps select optimal split versus packaged designs.

Central HVAC Systems

Central HVAC systems use one or more central plant equipment rooms to condition air and water distributed throughout the building. Cooling and heating capacities are designed to meet whole building loads. Central systems offer economies of scale but lack zoning flexibility. Common types:

- Central forced furnaces or air handling units supply cooled or heated air via ductwork.

- Hydronic or steam boilers generate hot water or steam sent to heat exchangers.

- Chillers produce chilled water circulated to air handlers or terminal coils.

- Rooftop units are packaged central systems for single-story structures.

Proper equipment sizing, distribution design, and control integration ensure central plant optimization.

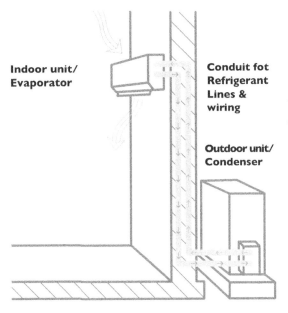

Mini-Split Systems

Mini-split systems provide zoning flexibility using one outdoor condenser/compressor serving multiple indoor evaporator units, each with dedicated thermostats and ductless air distribution. Multi-zone control improves comfort and efficiency. Concealed, compact, ductless indoor units allow flexible placement, including in retrofits. Smaller decentralized units reduce upfront costs. Mini-splits are common in residential additions and light commercial locations. Sizing the outdoor unit properly to meet the sum of indoor capacities is imperative.

Variable Refrigerant Flow (VRF)

Variable refrigerant flow (VRF) systems are a specialized mini-split system. The key differentiating feature is the ability to vary the refrigerant flow to each indoor evaporator based on individual zone loads. VRF systems dynamically redirect excess capacity to zones needing more cooling, significantly improving efficiency, comfort, and redundancy. Sophisticated controllers precisely modulate multiple compressors and valves in the outdoor unit and track individual indoor unit demands to enable this variable capacity zoning control. The flexibility and performance of VRF systems make them advantageous for large open floorplan offices and multifamily buildings.

Geothermal Heat Pumps

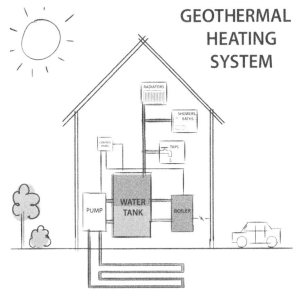

Geothermal heat pumps utilize subterranean earth temperatures to enhance heating and cooling efficiency. Systems circulate water or refrigerant through underground loops, transferring heat to and from the ground. The geothermal advantage stems from the earth's moderate year-round temperature. Cooling mode rejects heat absorbed from buildings into the warmer summer ground. Heating mode extracts relatively warmer winter ground heat for space warming. Though installation costs are high, drastically lower operating expenses offset long-term savings. Geothermal units can supply both forced air and hydronic distribution systems.

Chilled Beams

Chilled beam systems distribute cooling via convection from specialized induction or passive beam terminals. Chilled water flowing through each beam unit causes cool air to fall and hot air to rise, generating gentle convective air currents. This enables cooling delivery with minimal blower fan energy. Chilled beams couple well with radiant heating systems for a comfortable, quiet, draft-free indoor environment. Fan coils can supplement for occasional peak loads. Avoiding extensive ductwork reduces installation costs. Chilled beam systems are common in Europe and are gaining adoption globally.

Thermal Storage Systems

Thermal storage HVAC systems shift all or a portion of cooling/heating loads to off-peak hours by storing cold/hot water or ice for later use. This enables the downsizing of chillers and boilers by spreading peak demands over time. Stored cooling/heating is generated at night when utility rates are lower, and electricity is often cleaner. Thermal storage tanks improve resiliency during grid disruptions. These systems reduce operating expenses and environmental impacts when properly designed and controlled. Various storage media include water, ice, eutectic salts, aquifers, and the ground.

Radiant Heating and Cooling

Radiant systems distribute heating and cooling via thermal radiation and gentle convection from water-filled pipes or electric cables embedded in floors, walls, and ceilings throughout conditioned spaces. Radiant exchange directly heats/cools occupants and surfaces versus the air volume, as in forced air systems. This can enhance comfort while allowing wider setpoint ranges and lower airflow. Radiant systems couple well with ventilation units for fresh air needs. Designs avoid wasting energy conditioning unused zones. Radiant technology variations include hydronic tubing, heating cables, and thermally-activated surfaces.

Ventilation & Air Filtration Systems

Dedicated ventilation and air cleaning systems are vital for maintaining indoor air quality in tight, energy-efficient buildings. Exhaust ventilation fans remove stale air from bathrooms, kitchens, and utility rooms. Energy recovery ventilators employ heat exchangers to temper fresh incoming air using building exhaust airs. High-pressure air filters like HEPA units can effectively remove airborne particles and allergens when properly applied. Humidity control helps avoid mold risks. Understanding correct ventilation system implementation ensures air purity.

Customized & Integrated Systems

Hybrid configurations or fully customized HVAC systems tailored to specialized requirements may be justified for complex projects. Integrating complementary technologies like radiant cooling, chilled beams, desiccant dehumidification, heat recovery, thermal storage, advanced controls, and more can yield high-performance results. Simulations inform designs accounting for synergies. Large facilities often necessitate customized, optimized solutions balancing performance with practicality. Even small buildings can benefit from integrated thinking, such as combining mini-splits with ventilation units.

KEY COMPONENTS: IN-DEPTH ANALYSIS

HVAC systems comprise an interconnected web of components performing vital functions. Analyzing key components individually makes their specific operation and purpose clear, providing insights into keeping each one functioning reliably. Mastering the details of these pivotal pieces supplies the knowledge needed to diagnose issues and make modifications to improve efficiency.

Compressors

The compressor is the driving "heart" of HVAC refrigeration cycles, circulating and pressurizing gaseous refrigerant through successive phase change processes. However, conceptually simple devices, understanding compressor categories, configurations, controls, and maintenance factors is essential for proper operation. Positive displacement compressors like reciprocating and rotary designs increase pressure by mechanically reducing the volume occupied by the refrigerant vapor via pistons or rotating mechanisms. Dynamic compressors like centrifugal models utilize impellers to accelerate flow and induce pressure rise through velocity increase. Each category has relative advantages. Electric motors, gas/diesel engines, steam turbines, or solar Stirling engines can provide input power. Efficiency comparisons help select optimal energy sources for a given application. Two-stage compression with intercooling leverages multistage thermodynamic benefits. Variable frequency drives allow modulation to match load variations precisely, saving substantial energy compared to on/off control. Proper compressor selection, installation, control, and maintenance sustains cooling performance over years of demanding use. Maintaining routine lubrication, inspection, and protection device servicing prevents avoidable failures.

Heat Exchangers

Heat exchangers are vital HVAC components that transfer collected heat between fluids, refrigerants, and air streams safely and efficiently. They use conductive, convective, and radiant heat transfer across large surface areas separating the flows. Performance depends on optimal thermal design and material selection. Finned-tube construction enhances surface area and turbulence to boost heat transfer coefficients. Plate and frame exchangers provide high effectiveness through alternating fluid layers in compact, counterflow paths. Shell and tube designs allow easy inspection and cleaning of straight tubes but require greater material costs. Pipe coils offer simplicity at the expense of lower capacity per core volume. Each approach has merits. Common heat exchanger materials like aluminum, copper, and steel are chosen for high conductivity, corrosion resistance, and cost factors. Thermal resistance minimization and heat transfer maximization govern design details like small fluid passages, extended finned areas, and continuous, leak-free seals. Keeping heat transfer surfaces clear of fouling is imperative through treatment or proper maintenance procedures.

Fans and Blowers

Fans and blowers provide the motive force that propels conditioned air through HVAC systems and spaces via airflow creation. They convert input motor power into useful air movement. Matching fan static and dynamic lift capabilities to duct system resistance curves prevent excessive noise while moving the required air volumes. Fan laws help predict performance changes with adjustments to rotation speed, blade geometries, and other factors. Centrifugal blowers utilize a spinning impeller to impart velocity to air entering its eye for high-pressure differential applications like furnaces. Axial fans have rotor blades mounted parallel to airflow for lower-pressure ducted systems. Propeller fans move air along their axis and are used in localized ventilation. Fan arrays, variable pitch blades, and motor speed controls provide capacity modulation for efficiency. Careful fan selection, sizing, installation, and control integration ensure critical airflows are maintained while minimizing energy consumption across operating conditions. Quiet, properly sized HVAC fans improve comfort while lasting for years.

Pumps

Pumps circulate chilled or heated fluids in hydronic HVAC heating and cooling distribution systems by converting rotating mechanical energy into fluid flow and pressure rise. Proper pump sizing, selection, placement, and control deliver heating and cooling reliably and efficiently to all terminals. Centrifugal pumps use an impeller to accelerate liquid radially outward from the rotor shaft, pressurizing and discharging it. Positive displacement pumps like piston or diaphragm types induce flow via volume reduction in an enclosure. Matching pump curves to system resistance while maximizing efficiency is key.

Pump configurations for primary-secondary distribution, variable flow, pressure booster roles, and integrated controls affect overall performance. Careful installation and maintenance ensure proper anchoring, sealing, venting, lubrication, and bearing protection. Energy-saving measures like variable speed drives and staging based on demand optimize pump operation.

Condensers and Evaporators

Condensers and evaporators are specialized heat exchangers central to vapor compression refrigeration's phase transition processes. The condenser serves as the heat rejection component, releasing collected heat to ambient air or water to condense hot compressed gaseous refrigerant entering from the compressor into a liquid state.

The evaporator then absorbs ambient heat into the refrigerant liquid, causing it to boil and evaporate into a gas that can absorb substantial heat energy. Air or fluid flows transferring heat across the condenser and evaporator coils and fins are care-

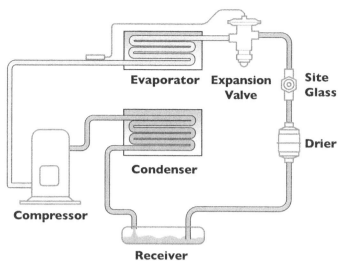

fully optimized along with thermodynamic cycle parameters to maximize efficiency. Keeping internal flow passages and external surfaces clear of fouling maintains performance. Proper condenser and evaporator selection, sizing, and installation matched to anticipated cooling loads and environmental factors are imperative to achieving design capacity targets and energy efficiency expectations. Their effectiveness and reliability directly impact overall system performance.

Thermostatic Expansion Valves

Thermostatic expansion valves (TXV) are metering devices that regulate the flow of liquid refrigerant entering the evaporator in response to temperature and pressure signals. This regulates the refrigerant available in the evaporator to match cooling loads. TXVs prevent evaporator flooding while avoiding starving. An internal diaphragm senses the evaporator outlet superheat and adjusts a tapered needle to control flow accordingly.

Bulbs filled with gas or liquid charge apply superheat-sensing pressure onto the diaphragm. External equalizer lines can reference evaporator pressure directly. TXV adjustments alter refrigerant flow rate, pressure drop, superheat, and capacity. Careful installation matching valve size and sensing bulb placement to the evaporator is critical. Properly functioning TXVs optimize vapor compression cycle efficiency.

Refrigerant Piping

Refrigerant piping carries hot pressurized gases between compressors and condensers and liquid refrigerant to evaporators. Proper materials, sizing, layout, joint connections, and thermal insulation are required to maintain refrigerant cycle integrity. Copper and steel with brazed, welded, flared, or threaded fittings make reliable seals able to contain high pressures for many years. Correct diameters balance friction losses against flow velocity limits. The shortest routing minimizes pressure drops. Gradual bends ease turbulence. Isolation valves enable section isolation for repairs. Insulation reduces surface condensation and heat transfer in or out of the pipes. Careful leak testing, pressure testing, and evacuation remove contaminants before refrigerant charging. Following best practices in piping and insulation boosts efficiency.

Electrical Controls

HVAC systems integrate extensive electrical controls to modulate process variables and provide automated, optimized operation in response to changing conditions and loads. Thermostats, humidistats, and other sensors provide critical input on indoor environmental conditions. Microprocessors running control algorithms like PID loops manipulate actuators to hit setpoints. Relays, contactors, and starters enable sophisticated staging and cycling of equipment like compressors and evaporator fans to match loads and maximize efficiency closely. Variable frequency motor drives also allow flow modulation. Dashboards and remote connectivity facilitate monitoring and analysis. Properly designing, adjusting, and maintaining HVAC integrated controls is imperative to performance.

Modern air conditioning, refrigeration, and HVAC systems utilize an extensive vocabulary of technical jargon and industry-specific terminology. Successfully navigating this complex, specialized language is critical for effective communication, properly implementing best practices, and fully comprehending educational resources.

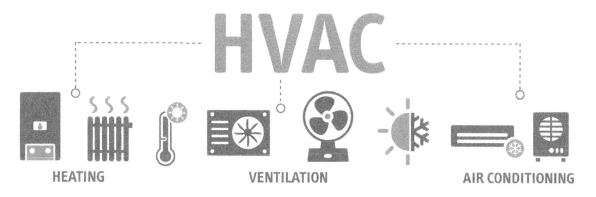

Thermodynamics Terms

Thermodynamics is the branch of physics that deals with the relationships between heat, work, energy, and temperature. HVAC and refrigeration deeply leverage fundamental thermodynamic principles and processes to achieve heating, cooling, and heat pumping.

- **Enthalpy** refers to the total heat content or thermal energy stored within a substance. It is expressed in BTUs per pound. The enthalpy of a refrigerant or air sample has important implications for HVAC performance analysis.

- **Entropy** is the degradation or loss of usable energy due to irreversibility and inefficiencies within any real-world process. High entropy processes discharge significant wasted heat. In HVAC systems, entropy increases represent declining efficiency over time or design flaws.

- **Sensible heat** changes a substance's temperature without changing its phase. It results in a temperature rise or decline. An example is a gas burner heating room air - the temperature increases, but its gas phase does not change.

- **Latent heat** is heat absorbed or released when a substance changes phase between liquid, solid, and gas states. This phase change latent heat is significant compared to sensible heat. Condensing and evaporating refrigerants utilize latent heat transfer.

- **Specific heat** is the energy required to raise one pound of a substance by 1° degree Fahrenheit. It varies substantially by material. Water's high specific heat makes it useful for thermal storage. Specific heat is an important parameter in HVAC load calculations.

- **Heat flux** is the heat transfer rate through an area over time. It is expressed in BTUs per square foot per hour. Understanding heat flows into and out of conditioned spaces allows proper sizing of HVAC equipment. High heat flux indicates good heat exchanger performance.

- **Thermal resistance** is the resistance to conductive heat flow through a material. It is the inverse of thermal conductance. Insulators have high thermal resistance, while metals demonstrate low resistance. Computing the total resistance of composite materials informs heat loss analysis.

- **Heat capacity** is the maximum amount of heat an object can absorb per degree of temperature rise. It is the product of the object's mass and specific heat capacity. Heat capacity determines how much energy is required to change an object's temperature.

- An **adiabatic process** is a thermodynamic change occurring within a system without transferring heat into or out of the working fluid. HVAC components like evaporator coils aim to approach adiabatic behavior but ultimately lose some heat to surroundings.

Refrigerant Terms

Refrigerants are specialized fluids circulating through vapor-compression cycles, enabling heat absorption and rejection through phase changes.

- **PSIG** stands for pounds per square inch gauge. It is the standard unit used for measuring refrigerant pressures within a system. Refrigerant pressures must remain within design limits for proper operation and safety.

- **Subcooling** refers to the number of degrees a liquid refrigerant is below its saturation point at a given pressure. Proper refrigerant subcooling indicates the condenser is operating effectively and prevents premature boiling in lines.

- **Superheat** is the number of degrees a refrigerant vapor exceeds its saturation temperature at a given pressure. Maintaining superheat out of the evaporator protects the compressor and indicates proper low-side operation.

- **The triple point** is the pressure and temperature where refrigerants can exist simultaneously as solid, liquid, and vapor states. Understanding refrigerant triple points ensures proper operation within pressure/temperature envelopes.

- **The critical point** is the maximum temperature and pressure of a refrigerant as a liquid. Beyond the critical point, it can only be a gas regardless of pressure changes. Each refrigerant has unique critical point values.

- A **zeotrope** is a refrigerant mixture where the component fluids have different boiling points. This allows a temperature glide rather than a single boiling point temperature during phase change. Zeotrope blends aim to enhance performance.

- An **azeotrope** is a refrigerant blend that evaporates and condenses at a single temperature like a pure fluid, with no temperature glide. Azeotropes attempt to mimic single-component refrigerants.

HVAC Equipment Terms

HVAC systems comprise complex machinery and components, each with industry-specific names.

- A **condenser** is a heat exchanger in which hot compressed refrigerant vapor rejects heat and condenses into a liquid. It facilitates heat rejection from the vapor compression cycle into outdoor ambient air or water.

- A **chiller** is a large piece of equipment that chills water, glycol, or another fluid to supply cooling capacity indirectly to a building rather than directly cooling air. Chillers serve central building plants.

- A **coil** assembles tightly spaced heat exchanger tubes shaped into a coil configuration to maximize surface contact area for heat transfer. Coils take several forms in HVAC systems, including cooling coils and condenser coils.

- A **compressor rack** is a framework for mounting multiple units with integrated refrigerant piping, controls, and accessories. Rack systems allow modular equipment room configurations.

- A **boiler** is equipment that heats water or a water mixture to supply building heating capacity, typically via the combustion of fuels like natural gas or electric resistance heating.

- A **furnace** is an indoor appliance that provides warm air heating from burning fuel, electric heating elements, or hot water coils. Furnaces distribute heat through ductwork or directly within a space.

- An **evaporative cooler** utilizes the evaporative cooling effect of water to produce cooled air. In dry climates, evaporative coolers offer a low-energy alternative to compressors for cooling.

HVAC Configuration Terms

Numerous designations are used to characterize HVAC system types and arrangements.

- **VAV** is a variable air volume HVAC system that modulates airflow to zones based on changing cooling or heating demands rather than maintaining a constant static flow rate. VAV reduces energy use.

- **Multi-zone systems** have multiple independently controlled zones, each with its temperature control through a dedicated thermostat or controller. This provides improved comfort and efficiency.

- An **airside economizer** uses cool outdoor air for free cooling when ambient conditions permit shutting down mechanical refrigeration compressors and saving energy. Airside economizers leverage "free" cooling.

- A **waterside economizer** uses cool tower water to provide free cooling to chillers and other equipment when cold enough, thereby saving chiller energy. It transfers heat to towers rather than refrigerating it.

- **Water-source heat pumps** utilize a common water loop to exchange heat rather than individual outdoor air coils. The water loop allows heat pumps to use centralized equipment like boilers and cooling towers.

HVAC Performance Terms

Key HVAC efficiency and capacity metrics help characterize equipment performance.

- **AFUE** stands for Annual Fuel Utilization Efficiency. This percentage represents the fuel converted to usable heat in heating systems like furnaces. A high AFUE indicates an efficient heating system.

- **SEER** stands for Seasonal Energy Efficiency Ratio. This metric rates air conditioners' and heat pumps' relative cooling energy performance over a year, accounting for seasonal variations. A higher SEER indicates more efficient equipment.

- **EER** stands for Energy Efficiency Ratio. This measures equipment cooling efficiency under standardized test conditions. EER facilitates performance comparisons between different model units.

- **COP** stands for Coefficient of Performance. The cooling or heating output ratio is divided by the electrical energy input. A higher COP equates to greater efficiency.

- **kW** per ton is the electrical power input needed by a cooling system in kilowatts divided by the cooling output achieved in tons. Lower kW/ton values indicate more efficient equipment.

- **Delta T** represents the temperature difference between two interacting fluid streams in a heat exchanger. The delta T impacts heat transfer performance and efficiency. A larger delta T implies greater potential heat transfer.

HVAC Distribution Terms

Ductwork, piping, and ancillary components distribute conditioned air and fluids through HVAC systems.

- **Supply air** is the cooled, heated, or ventilated air flowing to a space to meet target environmental conditions and loads. It represents the useful output of HVAC processes.

- **Return air** is recirculated from a conditioned space to the HVAC unit after circulation. It must be handled appropriately to achieve comfort and efficiency goals.

- A **plenum** is a specialized duct area merging multiple airflow paths from supply ducts into a shared common stream. Proper plenum design is important for balancing airflows.

- A **register** is a grille or vent covering the outlet end of an air supply duct that allows the conditioned air to enter a space. Registers can control airflow direction.

- A **diffuser** is a circular, square, or rectangular device installed at duct openings designed to spread out and mix airflow streams, eliminating concentrated high-velocity drafts.

- A **riser** is a vertical distribution pipe or duct that connects to different floors or sections of a building to allow conditioned air or water to reach multiple levels efficiently.

- A **VAV box** is a variable air volume box that modulates the amount of airflow through a duct run by opening or closing dampers based on localized temperature demands.

HVAC Operations Terms

Smooth ongoing operation of HVAC systems in buildings relies on some standard practices and concepts.

- A **setpoint** is the desired target temperature configured on a thermostat or building automation system controller. HVAC systems react to changes in conditioned variable values relative to the setpoint.

- **The dead band** refers to an acceptable range above and below a setpoint where no corrective action allows some fluctuation. Defining a dead band prevents excessive system cycling from overcorrection.

- An **energy management system (EMS)** is a comprehensive computerized system for monitoring and controlling equipment like HVAC systems across an entire facility for performance optimization. They Track trends and identify issues.

- **Building automation systems (BAS)** integrate control of HVAC, lighting, safety, and other critical building-wide systems on an automated networked platform for centralized monitoring and intelligent control.

- **Firmware** consists of low-level software instructions programmed into non-volatile memory chips integrated into system control hardware to enable basic operation. Most HVAC controls contain firmware.

- A **piping and instrumentation diagram (P&ID)** provides a detailed schematic visually mapping out the piping, valves, instrumentation, and control system components in a process system like a chiller plant.

HVAC Load and Performance Calculations

Properly sizing HVAC equipment and calculating heating and cooling loads utilizes key industry metrics and analysis methods.

- A ton represents a unit of cooling capacity defined as the heat energy transfer **needed** to melt one ton (2000 lbs) of ice over 24 hours, equivalent to removing 12,000 British thermal units (BTUs) per hour.

- **Square feet per ton** is the area that can be served or conditioned per ton of cooling or heating capacity in a system. It helps size equipment and ducts appropriately for a given space.

- **R-value** measures the thermal resistance or insulating effectiveness of a material or composite assembly like a wall or roof. Materials with higher R-values have lower heat conduction and greater resistance to heat flow.

- **U-factor** quantifies the inverse property, thermal conductance, indicating the heat transfer rate through a building envelope element in BTU per hour per square foot per degree F of temperature difference (BTU/hr/ft2/°F). A lower U-factor indicates superior insulation.

- **Infiltration** refers to the uncontrolled leakage of outdoor air into a building through small cracks and gaps in the facade, windows, doors, or roof. It is commonly measured in air changes per hour (ACH) relative to the building's volume. Higher infiltration represents greater heating and cooling loads.

- The **solar heat gain coefficient (SHGC)** represents the ratio of solar radiation admitted through a window to the radiation impacting the glass. Lower SHGC glass reduces solar heat gain.

HVAC Mechanical Equipment

HVAC systems incorporate extensive mechanical equipment, including compressors, pumps, valves, and actuators.

- **Head pressure** represents the high side pressure on the discharge side of a compressor in a refrigeration cycle. Head pressure must remain in proper range for efficient system operation and longevity of equipment.

- **Lift** refers to the increase in air or fluid pressure that a fan or pump must generate to overcome the flow resistance in an attached duct or piping system. Designing for proper lift prevents issues like insufficient airflow.

- **Static pressure** is the pressure imposed on an air stream by the friction and turbulence of air moving through ductwork, coils, filters, and other components that constitute resistance to airflow.

- **Velocity pressure** is the dynamic pressure related to the velocity of moving air that a fan or blower must overcome in addition to static pressure to deliver sufficient airflow. Fast-moving air has high-velocity pressure.

- **Belimo valves** are electromechanical control valves designed for HVAC systems that can be remotely positioned by modulating the voltage of an input control signal. This gives precision flow control without a large actuator required.

- A **thermal expansion valve (TEV)** is a refrigerant metering valve that automatically adjusts its opening degree based on sensing evaporator outlet conditions to regulate refrigerant flow and properly maintain optimal system operation.

HVAC Safety Essentials

Safe, reliable, long-term operation of HVAC systems relies on consistent safety procedures and hazard awareness.

- **Lockout/tagout** procedures involve physically isolating and locking out equipment from all hazardous energy sources during maintenance to prevent accidental startup and protect technicians.

- A **confined space** refers to enclosed areas large enough for bodily entry and work but with potentially hazardous conditions like poor ventilation, requiring special precautions and procedures to enter safely and avoid asphyxiation hazards.

- **Fall protection** refers to equipment like full-body harnesses, lanyards, and engineered anchor attachment points designed to catch and safely arrest the fall of workers exposed to fall hazards when working at heights to prevent severe injury or death. Proper training is critical.

- **Personal protective equipment (PPE)** includes gloves, respirators, hearing protection, eye shields, hard hats, and more that protect HVAC workers against workplace hazards. Correct PPE use is legally mandated.

HVAC Maintenance Essentials

Consistent performance and longevity of HVAC equipment requires diligent maintenance work and adherence to recommended schedules.

- **Preventive maintenance** involves performing regularly scheduled equipment servicing, inspection, and cleaning tasks to prevent avoidable failures, improve equipment lifetime, and sustain energy efficiency.

- **Air balancing** refers to adjusting dampers, valves, and other controls to fine-tune the distribution of design airflow and water flow rates through an HVAC system to meet initial commissioning parameters. Proper balancing maximizes comfort.

- **Duct leakage testing** involves specialized pressure measurements to quantify air leakage in installed ductwork using a blower door system or similar equipment. Leak testing ensures ducts deliver design airflows.

- **Insulation integrity** means verifying that ducts, pipes, and equipment insulation remains intact, undamaged, and appropriately sealed against infiltration or external moisture. Maintaining insulation integrity minimizes system losses.

HVAC Commissioning Concepts

Commissioning is the process that ensures HVAC systems function as designed from installation through occupancy.

- **Commissioning** includes thoroughly testing and validating the proper operation of all HVAC equipment, components, control programming, and integrated system performance before owner occupancy. Finding problems before turnover reduces future issues.

- **Functional performance testing** compares equipment operation to specifications across various conditions like peak capacity, part load, and minimum turndowns. Data verifies factory performance claims.

- **Validation** confirms a system meets all owner project requirements. It proves operation satisfies needs related to capacity, efficiency, comfort, maintainability, resilience, cost, and other parameters specified during planning.

- **Retro-commissioning** applies the commissioning process to existing systems to optimize performance, identify problems, and improve efficiency and reliability through tune-ups and upgrades. The process is often repeated periodically.

Electrical and Controls

Modern HVAC systems incorporate extensive electrical and electronic control technologies to provide functionality, precision, data access, and automated intelligent optimization.

- A **protocol** defines the standardized digital language, rules, and data formats that enable diverse devices and systems to communicate, share data, and interoperate on a network. Common HVAC protocols include BACnet, LonWorks, and Modbus.

- A **transformer** is an electrical component of two separate wire coils wound on a common core that is used to convert between high-voltage building electrical distribution and lower standard HVAC system control voltages through the principles of electromagnetic induction.

- A **solid-state** relay is an electronic switching device used for HVAC electrical system control that contains no moving parts, only solid-state electronics. This increases lifespan and switching speed compared to mechanical relays.

- **Proportional-integral-derivative (PID)** control is a control algorithm approach that utilizes feedback loops, tuning parameters, and input variables to automatically adjust some processes to hit and maintain a target setpoint value in equipment like variable air volume boxes.

- **Firmware** consists of operating software instructions permanently programmed into non-volatile read-only memory integrated circuits on circuit boards inside HVAC controls hardware. The firmware enables basic operation and control functionality.

- **Ladder logic** is a graphical programming language used for programmable logic controllers (PLCs) that represent circuits and controller actions resembling relay logic diagrams used to control industrial machines and processes. Ladder logic is commonly used in building automation systems.

HVAC Design and Analysis

HVAC systems' design and analysis process relies on specialized terminology and methodologies.

- A **heating load calculation** determines the maximum rate of heat loss from a building in peak design conditions, determining the required heating system capacity to maintain interior setpoint temperatures. Load depends on transmission, infiltration, and ventilation losses.

- A **cooling load calculation** analyzes the maximum anticipated rate of heat gain into a building in peak summer design conditions. Calculated loads dictate the required cooling capacity to maintain interior conditions. Heat gain components include solar gains, occupant loads, and equipment.

- **Psychrometrics** involves analyzing the thermodynamic properties of air and water vapor mixtures across varying conditions of temperature, humidity, enthalpy, and moisture content. Psychrometric charts model air conditions, which inform HVAC design.

- **Computational fluid dynamics (CFD)** uses numerical analysis and fluid flow modeling algorithms performed by specialized computer software to simulate and predict airflow, temperature distributions, and airside performance in complex spaces and HVAC components.

- **Sound transmission class (STC)** ratings measure the effectiveness of structures at blocking sound based on the transmission loss across standard frequencies. High STC designs prevent HVAC equipment noise issues.

- **Vibration isolation** refers to using specialized spring mounts, pads, bases, or vibration-dampening hangers to reduce the transfer of equipment vibrations produced by rotating components like motors or fans into the surrounding building structure to prevent noise problems or safety risks.

HVAC Heating Equipment

Heating systems involve particular equipment, methods, and metrics.

- A **forced air furnace** heats air using combustion or electric resistance coils and distributes it through ductwork to conditioned spaces. Furnaces offer zoned heating with moderate upfront costs.

- **Hydronic heating** uses boilers to heat water from piping to radiators, coils, or radiant tubing to transfer warmth. Avoiding extensive ducting cuts costs.

- **Heat pumps** use refrigeration cycles to absorb heat from outdoor or indoor air and pump the heat indoors for space and water heating, offering greater efficiency than alternatives.

- **Heat flux** refers to the rate of heat energy transfer through a given surface area over time. It is expressed in BTU per hour per square foot and used to analyze building envelope losses.

- **AFUE (Annual Fuel Utilization Efficiency)** measures the percentage of input fuel converted into useful heat output in thermal equipment like furnaces or boilers. Maximizing AFUE improves efficiency.

HVAC Cooling Equipment

Cooling systems likewise rely on specialized concepts, equipment, and metrics.

- **The vapor-compression cycle** is the thermodynamic refrigeration cycle circulated in a closed loop used in most cooling systems equipment like window air conditioners, split systems, and chillers.

- **Heat gain** is the cumulative rate of unwanted heat transferred into a cooled interior space from external and internal sources like solar gains, transmitted loads, and internal equipment. Heat gain is expressed in BTUs per hour.

- The **sensible heat ratio (SHR)** is the percentage of an HVAC system's total cooling load associated with lowering air dry bulb temperature instead of latent moisture removal. SHR helps size equipment.

- **Wet-bulb temperature** reflects the lowest temperature air can reach through evaporative cooling as water evaporates. It indicates the effect of humidity on perceived temperature.

- A **ton of cooling** refers to a unit of cooling capacity equivalent to melting one short ton (2000 lbs) of ice over 24 hours, corresponding to removing 12,000 British thermal units (BTUs) per hour.

- **SEER (Seasonal Energy Efficiency Ratio)** measures seasonal cooling equipment efficiency, including during peak summer conditions. Higher SEER values represent more efficient air conditioners and heat pumps.

HVAC Ventilation Essentials

Proper ventilation is vital for providing healthy indoor air quality and satisfying code minimum airflow requirements.

- **Air changes per hour (ACH)** measures the number of complete air volume replacements provided by mechanical ventilation systems over one hour. Higher ACH corresponds to greater fresh air dilution rates.

- **Exhaust ventilation** utilizes fans to remove stale or contaminated air from interior spaces like bathrooms and kitchens where moisture, odors, and pollutants are generated.

- **Energy recovery ventilators (ERVs)** exchange heat and moisture between outbound stale air and incoming fresh air using rotating heat exchangers to conserve energy.

- **Demand-controlled ventilation (DCV)** utilizes sensors and variable speed drives to dynamically modulate ventilation rates based on real-time occupancy,

CO2 levels, or other indicators instead of running fans constantly, saving energy.

- **Outdoor** air is external fresh air, typically filtered, that is purposefully brought into a building through HVAC equipment to replace stale interior air and properly ventilate occupied spaces as required by code.

HVAC Environments and Air Quality

Achieving occupant comfort and satisfaction requires properly controlling key indoor environmental factors.

- **Indoor air quality (IAQ)** refers to the cleanliness and healthiness of interior air as defined by multiple metrics, including temperature, humidity, particulates, VOCs, CO_2, and more. Poor IAQ can threaten occupant health, comfort, and productivity.
- **Carbon dioxide (CO_2)** is produced by building occupants through respiration. CO_2 concentration levels are monitored as a proxy for ventilation effectiveness and IAQ, indicating if adequate fresh air is being supplied.
- **Volatile organic compounds (VOCs)** are emissions from materials, paints, and processes that can accumulate indoors and be harmful at excessive levels. Limiting VOCs is an important consideration in finishing selections.
- **Particulates** are microscopic airborne particles and allergens ranging from dust to smoke that can cause respiratory issues if allowed to reach high indoor concentrations through poor filtration.
- **Humidity** describes the amount of water vapor in the air as a percentage relative to the maximum possible amount at that temperature. Too much or too little humidity causes comfort, health, and equipment issues.
- **Sick building syndrome** refers to negative occupant health outcomes like headaches, fatigue, and respiratory distress attributed primarily to time spent in buildings with poor indoor air quality and inadequate ventilation.

Building Automation Systems

Modern HVAC systems are increasingly integrated into building-wide automation systems and "smart" networks to provide monitoring, control, and data analytics.

- A **building automation system (BAS)** interconnects HVAC, lighting, safety, and other critical facility systems for centralized remote monitoring and automated intelligent control from a desktop or mobile interface.

- **Smart buildings** contain networked devices and sensors that allow automated optimization of operations like HVAC runtimes, setpoints, and schedules based on granular occupancy data, energy prices, weather forecasts, and other real-time inputs instead of static programming.

- The **Internet of Things (IoT)** refers to the network of internet-connected physical objects embedded with sensors, software, and controls that enable intelligent remote monitoring, data gathering, analysis, and adaptive control. HVAC systems integrate extensive IoT components.

- **Cloud computing** provides on-demand internet-based data storage, software services, and analytics via shared computing resources and infrastructure provisioned and managed remotely, allowing access from any connected device. Cloud-based building energy management platforms are increasingly common.

- **Fault detection diagnostics** utilize automated software algorithms to monitor system performance, detect faults or underperformance, diagnose likely causes based on analytical models, and guide technicians for repairs. FDD improves efficiency and uptime.

- This **expansive HVAC glossary,** compiling essential definitions in paragraph form under informative headings, empowers technicians, operators, and other professionals to navigate the complex terminology and jargon surrounding heating, ventilation, air conditioning, and refrigeration technology. The HVAC field encompasses various concepts, from fundamental science to advanced systems engineering. Thoroughly studying and mastering this applied vocabulary allows deeper engagement with the technology powering modern climate control.

CHAPTER 2

Installation and Maintenance Guidelines

STEP-BY-STEP HVAC INSTALLATION GUIDE

Installation of a heating, ventilation, and air conditioning (HVAC) system is a crucial process that demands precision, planning, and expertise. Whether you're a homeowner looking to tackle a do-it-yourself (DIY) installation project or a professional

HVAC technician seeking a comprehensive reference guide, this chapter, "Step-by-Step HVAC Installation Guide," will be your go-to resource.

Assessing Your Space: Sizing and Load Calculation

A crucial starting point in your HVAC installation journey is accurately assessing your space and determining the heating and cooling load required. Incorrect sizing can lead to inefficiency, energy waste, and discomfort. Proper load calculation is essential to ensure your HVAC system is the right size for your space.

a. Understanding the Importance of Proper Sizing

Proper sizing is a cornerstone of HVAC installation. If the system is too large, it will short cycle, wasting energy and reducing lifespan. If it's too small, it will struggle to maintain the desired temperature, leading to discomfort and higher energy bills. The goal is to find the Goldilocks solution – a just right system for your space.

b. Conducting a Load Calculation to Determine HVAC Capacity

A load calculation is a precise method of determining the heating and cooling capacity needed to maintain comfort in a space. It takes into account various factors, including:

- Climate: Local climate conditions play a significant role in determining the load your HVAC system must handle. The demands on your system will be much different in a cold northern climate compared to a hot and humid southern one.

- Insulation: The quality and quantity of insulation in your home greatly impact how much heating or cooling your system needs to provide.

- Occupancy: The number of people in your space also affects the load. A crowded room generates more heat and requires more cooling.

- Appliances: The heat generated by appliances, lighting, and other sources should be considered in the load calculation.

c. Factors Affecting Load Calculation

Understanding the factors that influence load calculation is vital to get an accurate result:

- Climate: The local climate data helps determine how much heating and cooling are necessary to maintain comfort.

- Insulation: The quality and quantity of insulation in your space significantly affect heat transfer.

- Occupancy: More people in a space means more body heat and increased cooling needs.

- Appliances: The heat generated by appliances, lighting, and other sources must be included in the calculation.

d. Using Industry-Standard Methods for Load Calculation

Load calculation is not an arbitrary process. It relies on industry-standard methods such as Manual J, developed by the Air Conditioning Contractors of America (ACCA). Manual J outlines a systematic approach to calculating heating and cooling loads. These calculations are complex but essential to finding the right-sized HVAC system for your space.

Manual J calculations consider heat loss and gain from various sources, making them a comprehensive and accurate way to size your system. They take into account factors such as:

- Building materials

- Insulation levels

- Window and door specifications

- Air infiltration rates

- Appliances and lighting

- Occupancy patterns

Using software or professional services to perform Manual J calculations ensures precision in your load calculation, which is fundamental for a successful HVAC installation.

Selecting the Right HVAC System: Factors to Consider

Choosing the right HVAC system is a decision that requires careful consideration. There's no one-size-fits-all solution, as various factors will influence the ideal system for your space. In this part of the chapter, we will walk you through the intricacies of selecting the right HVAC system:

a. Types of HVAC Systems

HVAC systems come in various types, and the one you choose should match your specific needs and the characteristics of your space. Here are the primary types to consider:

- Central HVAC Systems: Central systems consist of a central unit that provides heating and cooling to the entire space. They are common in larger homes and commercial buildings.

- Split Systems: Split systems are divided into outdoor and indoor units. These systems are suitable for smaller homes or single rooms.

- Window Units: Window units are designed to cool individual rooms and are affordable for small spaces.

- Ductless Mini-Split Systems: These systems offer zoned heating and cooling, making them ideal for areas with varying comfort needs.

- Packaged Systems: Packaged systems contain all components in one unit and are often used in smaller homes or spaces with limited installation options.

b. Matching System Types to Your Specific Needs and Space

The choice between these systems should be based on your space's characteristics and specific needs. Consider factors like:

- The size and layout of your space
- The number of rooms you want to condition
- Your heating and cooling preferences
- The local climate
- Energy efficiency goals

c. Energy Efficiency Ratings (SEER, EER, HSPF) and What They Mean

Energy efficiency is a critical aspect of HVAC system selection. The energy efficiency of a system is often rated using several metrics, including:

- Seasonal Energy Efficiency Ratio (SEER): SEER measures the cooling efficiency of an air conditioner or heat pump over a typical cooling season. A higher SEER rating indicates greater efficiency.

- Energy Efficiency Ratio (EER): EER is similar to SEER but measures a system's efficiency at a specific temperature and humidity level.

- Heating Seasonal Performance Factor (HSPF): HSPF measures the heating efficiency of heat pumps over a heating season. Like SEER, a higher HSPF rating indicates greater efficiency.

Understanding these ratings will help you choose a system that meets your energy efficiency goals and saves you money in the long run.

d. Considering Environmental Factors

Environmental considerations are increasingly important when selecting an HVAC system. It's crucial to be mindful of the following:

- Refrigerants: The type of refrigerant used in your system can impact its environmental footprint. Older refrigerants, like R-22, are being phased out due to their ozone-depleting properties. Newer, more environmentally friendly options are available.

- Eco-Friendliness: Some HVAC systems are designed to be more environmentally friendly. These systems may use energy-efficient technology, environmentally responsible refrigerants, or be part of a green building design.

- Government Regulations: Government regulations on HVAC systems constantly evolve to encourage eco-friendly choices. Staying informed about these regulations can help you make a greener choice.

e. Budget Considerations and Financing Options

HVAC system installation is a significant investment. It's essential to consider your budget and explore financing options if needed. Factors to consider include:

- The initial cost of the system
- Installation costs
- Potential utility bill savings
- Financing options, such as rebates, tax incentives, or low-interest loans
- Long-term maintenance and repair costs

Budget considerations should be balanced with the long-term benefits of a well-chosen, energy-efficient HVAC system. Investing in your comfort, energy savings, and the environment is an investment.

Ductwork Design and Installation: Ensuring Efficiency

Ductwork is the circulatory system of your HVAC installation. Proper design and installation of ducts are essential to ensure efficiency and comfort. In this section, we'll provide an in-depth look at ductwork:

a. The Importance of Well-Designed Duct Systems

Well-designed duct systems are essential for several reasons:

- Air Distribution: Ducts distribute heated or cooled air throughout your space. Proper design ensures even distribution, preventing hot or cold spots.
- Efficiency: Well-designed ducts minimize air leakage and heat loss, increasing the overall efficiency of your HVAC system.
- Comfort: Duct design can significantly impact the comfort of your space. Properly designed ducts help maintain consistent temperatures and humidity levels.

b. Choosing the Right Duct Material

Ducts are available in various materials, each with its advantages and disadvantages. The choice of duct material depends on factors like:

- Space requirements: The type of space where ducts will be installed, such as an attic, crawl space, or walls, will impact the choice of material.
- Insulation: Some duct materials provide built-in insulation, while others may require additional insulation for energy efficiency.
- Budget: The cost of duct materials varies, so budget considerations may influence your choice.

Common duct materials include sheet metal, flexible ducts, fiberglass duct boards, and ductless mini-split systems.

c. Proper Duct Sizing and Layout

The size and layout of your ducts are critical. Proper sizing ensures that your HVAC system can deliver the required amount of conditioned air to each room without straining the equipment or wasting energy.

- Calculating duct size: Duct sizing is based on factors like the airflow required, the duct length, and the airflow resistance. Software and duct design guides can help with these calculations.

- Layout considerations: The layout of ducts in your space should be designed to minimize bends, turns, and obstructions. Straight and well-supported ducts are more efficient.

- Zoning involves dividing your space into separate areas with thermostats and ducts. Zoning can increase efficiency and comfort by allowing you to condition different areas.

d. Sealing and Insulating Ducts to Prevent Energy Loss

Ducts can lose a significant amount of conditioned air through leaks and poor insulation. Proper sealing and insulation are crucial to reduce energy loss:

- Sealing ducts: Ducts should be sealed at joints and connections to prevent air leakage. Leaky ducts can significantly reduce system efficiency.

- Insulating ducts: Insulation helps maintain the air temperature inside the ducts. Uninsulated or poorly insulated ducts can lose or gain heat, reducing efficiency.

- Duct testing: After installation, ducts should be tested to ensure they are properly sealed and insulated. Duct leakage testing can identify and address any issues.

e. Testing and Balancing Your Ductwork

Testing and balancing your ductwork is essential to ensure that the system operates as intended:

- Airflow balancing: This process involves adjusting dampers or registers to ensure air is evenly distributed throughout the space. Balancing prevents hot or cold spots.

- Airflow measurement: Measuring the airflow in each branch of the duct system helps identify and correct imbalances.

- Pressure testing: Duct pressure testing verifies that the ducts are sealed properly and that there are no significant leaks.

Balancing and testing ensure that your HVAC system operates efficiently, maintains consistent comfort, and minimizes energy waste.

Electrical and Plumbing Requirements: A Comprehensive Overview

HVAC systems require both electrical and plumbing connections. In this section, we'll explore the electrical and plumbing considerations essential for a successful HVAC installation:

a. Electrical Requirements: Voltage, Circuits, and Dedicated Lines

HVAC systems have specific electrical requirements to function safely and effectively. These requirements include:

- Voltage: Understanding the voltage requirements of your HVAC system is essential. Most systems run on standard household voltages, but larger systems may require higher voltage.

- Dedicated circuits: HVAC systems should be connected to dedicated electrical circuits to ensure they can access the power they need without overloading other circuits.

- Safety precautions: Safety measures are crucial when working with electricity. Proper grounding, circuit protection, and adherence to electrical codes are essential.

b. Plumbing Connections: Drainage, Condensate Removal, and Water Supply

Many HVAC systems produce condensate, and this water must be properly managed:

- Drainage: Proper drainage is essential for removing condensate. Systems should be designed with drainage in mind to prevent water damage.

- Condensate removal: HVAC systems often include mechanisms for removing condensate, such as drain pans and pumps. Proper maintenance of these components is vital.

- Water supply: Some systems, like evaporative coolers, require a water supply. Ensuring that the water supply is clean and properly connected is crucial.

c. Safety Precautions During Electrical and Plumbing Work

Safety is paramount when working with electricity and plumbing during HVAC installation:

- Personal protective equipment (PPE): Proper PPE, including gloves, safety glasses, and insulated tools, should be worn to protect against electrical and plumbing hazards.

- Lockout/tagout procedures ensure electrical circuits are de-energized before working on them.
- Adherence to local codes: Following local building and safety codes is essential to ensure that electrical and plumbing work is done safely and legally.

d. Compliance with Local Building Codes and Regulations

HVAC installation must comply with local building codes and regulations. These codes cover a wide range of requirements, including:

- Electrical: Codes dictate how electrical work must be done to ensure safety.
- Plumbing: Plumbing codes cover installing drainage systems and water supply connections.
- Structural: Codes may dictate how HVAC equipment should be supported and installed within the building.
- Environmental: Some codes address environmental concerns, such as refrigerant handling and disposal.

Non-compliance with local codes can result in costly repairs and legal consequences. Working with a professional HVAC installer familiar with local codes is often the best way to ensure compliance.

Safety Precautions During Installation: Protecting Your Investment

Safety should always be a top priority during HVAC installation. Ensuring the process is carried out safely protects you and safeguards your investment in the HVAC system. In this section, we will explore key safety measures and best practices:

a. Personal Protective Equipment (PPE)

Personal protective equipment is crucial when working with HVAC systems. It provides a layer of protection against potential hazards. Some essential PPE includes:

- Safety glasses: Protect your eyes from debris, sparks, and chemicals.
- Gloves: Insulated gloves can protect your hands when working with electrical components.

- Respirator: Protect against dust and fumes that may be present during installation.
- Hard hat: Safeguard against falling objects if working in an area with overhead hazards.

Select the appropriate PPE for the specific tasks you perform during installation to reduce the risk of accidents and injuries.

b. Electrical Safety Measures

Working with electricity requires extreme caution. To ensure safety during installation:

- Turn off power: Disconnect the power source before performing any electrical work. Use lockout/tagout procedures to ensure the circuit remains de-energized while you work.
- Insulated tools: Use insulated tools designed for electrical work to reduce the risk of electrical shock.
- Avoid water: Keep electrical components away from water to prevent short circuits.
- Voltage test: Always verify that a circuit is de-energized using a voltage tester before touching any wires.

c. Handling Refrigerants Safely

Refrigerants are essential components of HVAC systems, but they must be handled with care:

- EPA certification: HVAC technicians must be certified by the Environmental Protection Agency (EPA) to handle refrigerants. Proper certification ensures the safe handling, recovery, and disposal of refrigerants.
- Leak detection: Regularly inspect the HVAC system for refrigerant leaks. Leaks not only affect system performance but can also release harmful refrigerants into the atmosphere.
- Recovery and recycling: Recover refrigerants during system maintenance and replacement and ensure they are properly recycled or disposed of according to environmental regulations.

d. Fire and Carbon Monoxide Safety

HVAC systems can pose fire and carbon monoxide risks if not properly installed and maintained:

- Combustion safety: Ensure that combustion appliances, such as furnaces, are properly vented to prevent the buildup of carbon monoxide.
- Smoke and carbon monoxide detectors: Install detectors in areas where combustion appliances are present to provide early warnings in case of leaks or malfunctions.

e. Ensuring Proper Ventilation During Installation

Ventilation is crucial when working with HVAC systems to prevent exposure to harmful fumes and ensure a safe working environment. Adequate ventilation is essential when:

- Soldering or brazing: The process of joining components through heat can release fumes that may be harmful if inhaled.
- Using adhesives or sealants: These products may produce strong odors and fumes that should not be inhaled.

Adequate ventilation is also important for dissipating heat when working in confined spaces.

Commissioning and Testing: Verifying System Performance

Installing an HVAC system doesn't end with connecting the components and securing the ductwork. It's essential to ensure that the system operates as intended. In this final section of the chapter, we will guide you through the commissioning and testing process:

a. Start-Up Procedures: Testing Components and Verifying Connections

The start-up phase is where your HVAC system begins to come to life. During this stage, you should:

- Verify electrical connections: Ensure all electrical connections are secure and properly wired.

- Test components: Test components like the blower motor, fan, and compressor to ensure they function correctly.
- Verify settings: Double-check the system's settings, such as thermostat programming, to ensure the system responds as expected.

b. Refrigerant Charge and Pressure Checks

The proper refrigerant charge is essential for system efficiency and performance. During commissioning, it's crucial to:

- Measure refrigerant levels: Verify that the refrigerant charge is correct, as specified by the manufacturer's guidelines.
- Check for leaks: Use electronic leak detectors or other methods for refrigerant leaks.
- Pressure checks: Measure refrigerant pressures in both the high and low sides of the system to ensure they are within the specified ranges.

c. Airflow Balancing and Performance Testing

Balancing the airflow and conducting performance tests are essential for ensuring even comfort and efficient operation:

- Airflow balancing: Adjust dampers and registers to achieve balanced airflow throughout the space. This prevents hot or cold spots.
- Performance testing: Test the system under various conditions, such as different temperature setpoints and fan speeds, to ensure it meets performance specifications.

d. Safety Checks and System Calibration

Safety should always be a priority. Perform safety checks and system calibration to ensure your HVAC system operates safely and efficiently:

- Safety checks: Verify that safety features like high-temperature limits and pressure switches are functioning correctly.
- System calibration: Fine-tune system settings, such as temperature differentials and cycling times, to optimize efficiency and comfort.

e. Handing Over the System to the End User

Once you've completed all necessary commissioning and testing, it's time to hand over the HVAC system to the end user, whether you or a client. Ensure that the user is well-informed about the following:

- System operation: Provide a thorough overview of operating the system, including setting the thermostat, changing filters, and addressing common issues.

- Maintenance requirements: Explain the essential maintenance tasks that must be performed regularly, such as cleaning filters and checking refrigerant levels.

- Contact information: Offer contact information for ongoing maintenance and support, ensuring that users can reach out if they encounter issues or have questions.

COMPREHENSIVE ROUTINE MAINTENANCE FOR PEAK PERFORMANCE

This chapter embodies the essence of keeping your HVAC system running smoothly, efficiently, and cost-effectively. Through monthly filter changes, cleaning and lubrication of moving parts, coil cleaning, refrigerant level checks, inspection of electrical components, and calibrating thermostats, we will equip you with the knowledge and skills to maintain your HVAC system at its best.

Monthly Filter Changes: Choosing the Right Filters

Your HVAC system relies heavily on the air filters to maintain indoor air quality and the system's efficiency. Regular filter changes are a simple yet crucial aspect of HVAC maintenance. In this section, we will discuss the significance of changing filters monthly and provide insights into selecting the right filters for your system.

Why Monthly Filter Changes Matter

Air filters are the unsung heroes of your HVAC system. They capture dust, debris, allergens, and other particles, preventing them from circulating through your home. Over time, filters become clogged and less effective, resulting in several issues:

1. Reduced Air Quality: Clogged filters can no longer effectively filter out particles, reducing indoor air quality. This can trigger allergies and respiratory problems for occupants.

2. Decreased Efficiency: When air filters are clogged, your HVAC system has to work harder to maintain the desired temperature. This extra strain not only reduces efficiency but also increases energy consumption.

3. Increased Repair Costs: The added strain on your system can lead to premature wear and tear, resulting in more frequent breakdowns and repair costs.

4. Shortened Lifespan: Regularly changing air filters is one of the most effective ways to prolong the lifespan of your HVAC system.

Selecting the Right Filters

Choosing the right filters for your HVAC system is as important as regularly changing them. Various types of filters are available, each with unique advantages and disadvantages. Here's a brief overview of common filter types:

1. Fiberglass Filters: These are inexpensive and capture larger particles but may not be as effective at trapping smaller particles and allergens.

2. Pleated Filters: Pleated filters offer better filtration than fiberglass filters and are reasonably priced. They are a popular choice for many HVAC systems.

3. HEPA Filters: High-efficiency particulate Air (HEPA) filters are known for their exceptional filtration capabilities. They can capture even the tiniest particles, making them ideal for individuals with allergies or respiratory issues.

4. Electrostatic Filters: These filters use an electrostatic charge to attract and capture

particles. They are more effective than standard fiberglass filters but may cost more.

5. Washable Filters: These filters are reusable, making them environmentally friendly. However, they require regular cleaning and may be less effective than disposable filters.

When selecting a filter, consider factors such as your budget, the needs of your household, and the recommendations of your HVAC system's manufacturer. Always follow the manufacturer's guidelines for filter replacement and maintenance.

Cleaning and Lubricating Moving Parts: Fan Blades, Bearings, and More

Your HVAC system consists of various moving parts, including fan blades and bearings, that need regular cleaning and lubrication to ensure smooth operation. This section will explore the importance of maintaining these components and provide step-by-step instructions for cleaning and lubrication.

Importance of Cleaning and Lubrication

Moving parts in your HVAC system, such as fan blades and bearings, play a crucial role in the system's operation. Over time, dust, dirt, and debris can accumulate on these components, leading to several problems:

1. Increased Friction: Accumulated debris can create friction, causing excessive wear and tear on moving parts.

2. Reduced Efficiency: As friction increases, the system's efficiency decreases, resulting in higher energy consumption and reduced cooling or heating performance.

3. Noise and Vibrations: Dirty or unlubricated parts can lead to unusual noises and vibrations, which are annoying and indicate potential issues.

4. System Breakdown: Neglecting cleaning and lubrication can eventually lead to the breakdown of vital components, resulting in costly repairs.

Cleaning Fan Blades

Cleaning fan blades is a relatively simple process, and it can significantly improve the efficiency and longevity of your HVAC system. Follow these steps:

1. Turn Off Power: Always ensure the power to your HVAC system is turned off before attempting any maintenance.

2. Access the Fan: Open the access panel to reach the fan blades. This is typically located in the indoor unit of your system.

3. Remove Debris: Carefully remove debris or dirt from the blades using a soft brush or vacuum cleaner. Be gentle to avoid damaging the blades.

4. Clean Blades: Mix a mild detergent with water and use a cloth or sponge to clean the blades. Ensure that the blades are completely dry before reassembling.

Lubricating Bearings

Bearings are essential components that allow various parts of your HVAC system to move smoothly. Proper lubrication ensures they function optimally. Follow these steps to lubricate bearings:

1. Locate Bearings: Identify the bearings in your system. They are typically found near the motor and fan blades.

2. Choose the Right Lubricant: Consult your HVAC system's manual to determine the recommended lubricant. Common options include oil or grease.

3. Apply Lubricant: Carefully apply the lubricant to the bearings according to the manufacturer's recommendations. Avoid over-lubricating, as this can be as detrimental as not lubricating.

4. Check for Smooth Operation: After lubrication, ensure the moving parts move smoothly without any resistance.

Regular cleaning and lubrication of moving parts in your HVAC system can prevent costly repairs and maintain the efficiency of your system.

Coil Cleaning: Evaporator and Condenser Coils

Coil cleaning is a critical aspect of HVAC maintenance that homeowners often overlook. This section will discuss the importance of cleaning the evaporator and condenser coils, along with step-by-step instructions for a thorough cleaning.

Why Coil Cleaning Matters?

The evaporator and condenser coils are vital components of your HVAC system and are responsible for heat exchange. Over time, these coils can accumulate dirt, dust, and debris, reducing their efficiency and causing several issues:

1. Reduced Cooling and Heating Efficiency: Dirty coils can't exchange heat effectively, reducing cooling and heating capacity.

2. Increased Energy Consumption: When the system's efficiency drops, it works harder to maintain the desired temperature, resulting in higher energy bills.

3. Compressor Strain: As the coils become clogged, the compressor is under increased strain, leading to potential breakdowns.

4. Shortened System Lifespan: Neglecting coil cleaning can lead to premature system failure.

Cleaning the Evaporator Coil

Cleaning the evaporator coil is a relatively straightforward process that can significantly improve your HVAC system's performance. Here's how to do it:

1. Turn Off Power: Ensure that the power to your HVAC system is turned off before starting any maintenance.
2. Locate the Evaporator Coil: The evaporator coil is usually inside the indoor unit. Access the coil by removing the access panel.
3. Remove Debris: Gently brush off any loose debris on the coil using a soft brush or a vacuum cleaner with a brush attachment.
4. Clean with a Coil Cleaner: Use a commercial coil cleaner or a mixture of equal parts water and vinegar to clean the coil. Apply the cleaner with a soft brush or a spray bottle.
5. Rinse and Dry: After cleaning, rinse the coil with water to remove any remaining cleaner. Allow the coil to dry completely before reassembling.

Cleaning the Condenser Coil

The condenser coil is typically located in the outdoor unit of your HVAC system. Cleaning is essential for optimal performance. Follow these steps:

1. Turn Off Power: As always, turn off the power to your system to ensure safety during maintenance.
2. Access the Condenser Coil: Open the access panel of the outdoor unit to reach the condenser coil.
3. Remove Debris: Clear leaves, dirt, or debris from the coil using a soft brush or a vacuum cleaner.
4. Clean with a Coil Cleaner: Apply a commercial coil cleaner or a mixture of water and vinegar to the condenser coil. Use a soft brush or a spray bottle to apply the cleaner.
5. Rinse Thoroughly: Rinse the coil thoroughly with a hose to remove any remaining cleaner and debris.
6. Straighten Coil Fins: Inspect the coil fins for any bending or damage. Use a fin comb to straighten any bent fins, ensuring optimal airflow.

7. Allow to Dry: Let the condenser coil air dry before reassembling the unit.

Regular cleaning of the evaporator and condenser coils is essential for maintaining the efficiency and performance of your HVAC system.

Refrigerant Level Checks: Avoiding Overheating or Freezing

The refrigerant in your HVAC system absorbs and releases heat, allowing your system to cool or heat your home. Maintaining the correct refrigerant level is crucial for system performance.

Why do Refrigerant Level Checks Matter?

Refrigerant is the lifeblood of your HVAC system, and having the correct level is essential for its proper function. Issues related to refrigerant levels can lead to a range of problems:

1. Inefficient Cooling or Heating: Too little or too much refrigerant can result in your system not being able to cool or heat your home effectively.

2. Increased Energy Consumption: Incorrect refrigerant levels force the system to work harder, increasing energy consumption and raising utility bills.

3. Compressor Damage: Inadequate refrigerant levels can cause the compressor to overheat, leading to potential breakdowns.

4. Reduced Lifespan: Neglecting refrigerant levels can lead to premature system failure.

Checking Refrigerant Levels

A qualified HVAC technician should perform refrigerant-level checks. It involves the following steps:

1. Pressure Test: The technician will attach pressure gauges to the system's low and high-pressure sides to measure the refrigerant's pressure.

2. Temperature Test: The technician will also measure the temperature of the refrigerant lines. This helps in determining if the refrigerant level is correct.

3. Refrigerant Addition or Removal: If the levels are incorrect, the technician will add or remove refrigerant to achieve the right balance.

Refrigerant level checks should be conducted during routine maintenance visits by a professional to ensure the optimal performance of your HVAC system.

Inspection of Electrical Components: Wiring, Capacitors, and Contacts

The electrical components of your HVAC system are vital for its operation, and ensuring they are in good condition is essential for preventing breakdowns and maintaining efficiency.

Why Inspecting Electrical Components Matters

Your HVAC system relies on a complex network of electrical components to function correctly. Neglecting their maintenance and inspection can lead to a variety of problems:

1. Breakdowns: Faulty wiring or damaged electrical components can lead to system breakdowns and costly repairs.

2. Reduced Efficiency: Electrical issues can hinder the system's performance, resulting in higher energy bills and decreased cooling or heating capacity.

3. Safety Concerns: Damaged wiring and components, such as electrical fires, can pose safety hazards.

The wiring in your HVAC system should be regularly inspected to identify any loose connections or damaged wires. Here's how to inspect the wiring:

1. Turn Off Power: Always turn off the power to your system before inspecting the wiring to ensure safety.

2. Visual Inspection: Carefully examine the wiring for any signs of damage, such as frayed or exposed wires.

3. Tighten Connections: If you find loose connections, tighten them with a wrench or screwdriver. Ensure that connections are secure but not over-tightened.

4. Replace Damaged Wires: If you discover damaged wires, replace them with wires of the same gauge and quality.

Inspecting Capacitors

Capacitors are crucial components that store electrical energy and boost the compressor and fan motors. Inspecting capacitors is vital for preventing system breakdowns. Here's how to inspect capacitors:

1. Turn Off Power: Ensure that the power to your system is off before beginning the inspection.

2. Visual Inspection: Examine the capacitors for signs of bulging, leakage, or damage. Damaged capacitors should be replaced.

Inspecting Contacts

Contacts control the flow of electricity in your HVAC system, and ensuring they are in good condition is essential. Here's how to inspect contacts:

1. Turn Off Power: As always, turn off the power to your system before inspecting contacts.

2. Visual Inspection: Carefully examine the contacts for any signs of pitting or burning. Damaged contacts should be replaced.

Regular inspection and maintenance of your HVAC system's electrical components can prevent breakdowns and ensure safe and efficient operation.

Calibrating Thermostats: Accuracy Matters

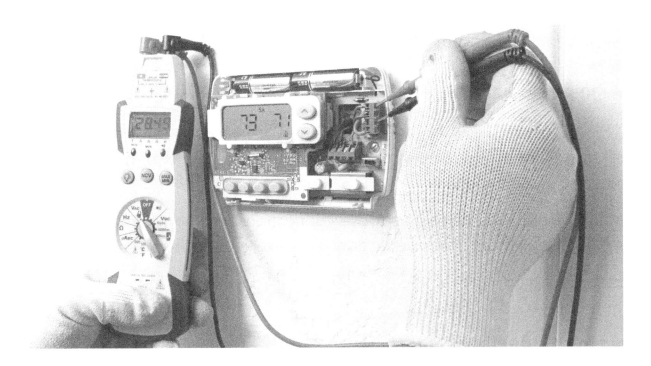

The thermostat is the control center of your HVAC system, and its accuracy is crucial for maintaining a comfortable indoor environment and reducing energy consumption.

Why Calibrating Thermostats Matters

The thermostat's primary function is to accurately measure the temperature in your home and signal the HVAC system to maintain the desired temperature. Inaccurate thermostats can lead to various issues:

1. Inconsistent Temperatures: Inaccurate thermostats may result in inconsistent temperatures throughout your home, making some rooms too hot or cold.
2. Higher Energy Bills: A thermostat that doesn't accurately reflect the indoor temperature can cause the system to run longer than necessary, increasing energy consumption.
3. Reduced Comfort: Inaccurate thermostats can lead to discomfort and a less-than-ideal indoor environment.

Calibrating Thermostats

Calibrating a thermostat involves adjusting it to measure the indoor temperature accurately. Here's how to calibrate a thermostat:

1. Use a Secondary Thermometer: Place a secondary thermometer in a central location in your home, away from direct sunlight or drafts. This thermometer will serve as a reference point.
2. Set the Thermostat: Set your thermostat to a specific temperature, such as 70°F (21°C). Give it time to stabilize.
3. Compare Readings: After the thermostat and secondary thermometer have stabilized, compare their readings. The thermostat may require calibration if there is a significant difference (more than a degree or two).
4. Calibrating Digital Thermostats: Digital thermostats can often be calibrated by following the manufacturer's instructions in the user manual. This typically involves accessing a calibration menu and making the necessary adjustments.
5. Calibrating Mechanical Thermostats: Mechanical thermostats can be calibrated by adjusting the anticipator setting. This setting determines how long the thermostat keeps the system running after reaching the desired temperature. Follow the manufacturer's instructions for making adjustments.
6. Repeat the Process: If the temperature reading is still inaccurate, repeat the calibration process until you achieve accurate temperature control.

Calibrating your thermostat accurately reflects the indoor temperature, allowing your HVAC system to operate efficiently and maintain a comfortable environment.

KEEP YOUR HVAC IN TOP SHAPE

INCREASE ITS LIFE & EFFICIENCY WITH THESE SIMPLE STEPS

#1 - NOW
- ○ Buy a high-efficiency pleated air filter.
- ○ Keep AC & heat units free of leaves, pollen, & grass.
- ○ Clear 2 feet of space around outdoor AC & heat units.

#2 - MONTHLY OR SEASONALLY
- ○ Inspect refrigerant lines each month.
- ○ Replace your air filter at least every 90 days.
- ○ Summer, turn off water to the furnace humidifier.
- ○ Fall, replace the humidifier filter and turn on water.

#3 - ANNUALLY
- ○ Replace your carbon monoxide detector battery.
- ○ Ensure outdoor AC unit is on firm and level ground.
- ○ Clear AC condensate drain with a bleach/water mixture.

SEASONAL MAINTENANCE CHECKLIST

Regular maintenance is the key to longevity and efficiency in HVAC systems. Like any other complex machinery, your HVAC system needs attention and care to function optimally. Adhering to a seasonal maintenance checklist ensures that your system operates smoothly throughout the year.

Spring Startup: Preparing for Cooling Season

As winter transitions to spring, it's time to prepare your HVAC system for the upcoming cooling season. Proper spring startup is essential to ensure that your air conditioning works efficiently and doesn't give you any unexpected troubles when the mercury rises. Following the steps in this section will make you well on your way to a comfortable and cost-effective cooling season.

Condensate Drain Inspection and Cleaning

One of the often-overlooked components of your HVAC system is the condensate drain. This small but vital part of the system removes excess moisture that results from the cooling process. Over time, dirt and debris can clog the drain, leading to issues like water leakage or reduced cooling efficiency. Here's how you can inspect and clean it:

1. **Inspect the drainpipe for clogs or blockages**: Begin by visually inspecting the drainpipe. Look for any visible obstructions that may be restricting water flow. Common culprits include dirt, algae, or debris buildup.

2. **Clear any debris**: If you find any obstructions in the drainpipe, clear them away. A simple, straightened coat hanger or a pipe brush can remove debris carefully. Be gentle to avoid damaging the pipe.

3. **Clean the condensate pan and drain line**: The condensate pan is underneath your

indoor unit's evaporator coil. Over time, this pan can accumulate algae or mold, which can obstruct the flow of water. Clean the pan using a mixture of warm water and a mild detergent. Ensure that the drain line is free from blockages as well.

4. **Ensure proper slope for drainage:** Check that the condensate drain line is correctly sloped to allow for proper drainage. It should slope away from the indoor unit and towards a drain or an outside area.

Checking Refrigerant Levels and Leaks

Refrigerant is the lifeblood of your air conditioning system. It circulates through the system, absorbing and releasing heat to keep your space cool. Over time, refrigerant levels can drop due to minor leaks, reducing the system's cooling capacity. To address this, you should:

1. **Check the refrigerant lines for any visible leaks or damage:** Inspect the refrigerant lines that connect the indoor and outdoor units. Look for any signs of oil stains, bubbling, or hissing sounds, which may indicate refrigerant leaks. If you find any of these signs, it's essential to have a professional technician locate and repair the leak.

2. **Use a refrigerant pressure gauge to measure levels:** You can use a refrigerant pressure gauge to check the refrigerant levels. Attach the gauge to the service port on the outdoor unit and compare the reading to the manufacturer's specifications for your specific system. If the pressure exceeds the recommended range, it indicates a refrigerant issue.

3. **Address refrigerant leaks promptly:** If you suspect a refrigerant leak, it's vital to contact a professional HVAC technician. Repairing refrigerant leaks requires specialized knowledge and equipment, and handling refrigerant can be hazardous. A professional will identify the leak's source, fix it, and recharge your system to the correct refrigerant level.

Testing the Thermostat and Fan Settings

Your thermostat is the control center of your HVAC system, allowing you to set the desired temperature and fan settings. Testing your thermostat and fan settings in the spring ensures that your cooling system functions efficiently and keeps you comfortable. Here's how to do it:

1. **Please turn on the thermostat and set it to a lower temperature than the

current room temperature. Start by turning on your thermostat and setting it to a temperature lower than the current room. This action should trigger the cooling system to start.

2. **Listen for the system to kick on and feel for cool air coming from the vents.** Pay attention to the sounds of your HVAC system. You should hear the indoor fan and the outdoor condenser unit turning on. Additionally, you should feel cool air coming from the vents.

3. **Ensure the fan is set to "Auto" rather than "On.":** It's important to check the fan settings on your thermostat. For energy efficiency, it's best to keep the fan in "Auto" mode, which means it will only run when the HVAC system is actively cooling the air. Running the fan continuously, in the "On" mode, can increase your energy usage without providing additional comfort.

Fall Shutdown: Getting Ready for Heating Season

As the cooling season ends and fall ushers in cooler temperatures, it's time to prepare your HVAC system for the heating season. Proper fall shutdown procedures can help you avoid unexpected issues when you switch from cooling to heating.

Cleaning and Inspecting the Heat Exchanger

The heat exchanger is a critical component of your furnace. It's responsible for transferring heat to the air circulating through your home. Over time, the heat exchanger can accumulate soot and debris, potentially becoming a safety hazard. Here's how you can clean and inspect it:

1. **Please turn off the furnace's power and allow it to cool down.** Before performing any work on your furnace, it's essential to turn off the power and allow the system to cool down completely. This step is crucial for safety reasons.

2. **Remove the access panel to access the heat exchanger:** Depending on your furnace's design, you may need to remove an access panel or open the combustion chamber to access the heat exchanger. Refer to your furnace's user manual for specific instructions.

3. **Gently brush away any soot or debris:** Once you can access the heat exchanger, use a soft brush or a vacuum cleaner with a brush attachment to remove any soot or debris. Be gentle to avoid damaging the heat exchanger's delicate surfaces.

4. **Inspect the heat exchanger for cracks or corrosion:** After cleaning, closely inspect the heat exchanger for any signs of damage, such as cracks, corrosion, or holes. If you notice any issues, addressing them promptly is crucial, as a damaged heat exchanger can release harmful gases into your home.

Lubricating Motors and Bearings

Your furnace contains motors and bearings that require lubrication to function properly. Neglecting this maintenance task can increase friction, wear, and motor failure. To keep your system running smoothly, follow these steps:

1. **Refer to your furnace's user manual to identify the motor and bearing lubrication points:** Different furnace models have various lubrication points. Refer to the user manual or the manufacturer's instructions to identify the locations requiring lubrication.

2. **Apply a few drops of oil to each lubrication point as the manual recommends.** Use the type of lubricant specified for each motor and bearing. Apply a few drops of oil to each lubrication point. Avoid over-lubricating, as excess oil can attract dust and debris.

3. **Use the appropriate lubricant for the specific motor or bearing type:** It's important to use the right lubricant as the manufacturer recommends. Applying the wrong lubricant can cause damage and lead to motor or bearing failure.

Testing the Ignition System (for Gas Furnaces)

If you have a gas furnace, the ignition system is crucial for safely and efficiently heating your home. Regular testing ensures that your furnace ignites properly and poses no safety risks. Here's how to test the ignition system:

1. **Turn up the thermostat to trigger the furnace to start:** Begin by turning up the thermostat to a temperature that should trigger the furnace to start. This action prompts the system to go through its ignition sequence.

2. **Listen for the sound of the ignition system sparking or glowing:** As the furnace starts, listen for the distinct sounds of the ignition system. You may hear a sparking sound or a glowing igniter, depending on your furnace type.

3. **Ensure that the burners ignite and warm air begins to circulate.** After the ignition sequence, the burners should ignite, and warm air should start circulating through your ducts. Confirm that these steps occur as expected.

If you encounter any issues with the ignition system during the test, it's important to address them promptly. An unreliable ignition system can lead to heating system failures and potential safety concerns. For gas furnaces, safety is of utmost importance. Inefficient ignition or issues with the burners can release harmful gases like carbon monoxide, which poses significant health risks.

TROUBLESHOOTING COMMON HVAC PROBLEMS

HVAC (Heating, Ventilation, and Air Conditioning) systems are pivotal in home comfort and energy efficiency. These complex systems are designed to maintain your living space's ideal temperature, humidity, and air quality, ensuring year-round comfort. However, like any intricate machinery, HVAC systems can encounter issues, and when they do, understanding how to diagnose and troubleshoot these problems is crucial.

Inadequate Cooling or Heating: Potential Causes and Solutions

Causes:

1. **Clogged Air Filters:** In the world of HVAC, one of the most frequent culprits behind inadequate cooling or heating is dirty or clogged air filters. These filters capture dust, pollen, and other airborne particles to maintain air quality. However, when they become congested with debris, they obstruct the airflow, forcing your HVAC system to work harder and less efficiently.

2. **Refrigerant Issues:** Low refrigerant levels can also result in inadequate cooling. A refrigerant leak or improper charging can reduce cooling or heating capacity.

3. **Faulty Thermostat:** If your thermostat isn't functioning correctly, it may not communicate with your HVAC system effectively, leading to temperature inconsistencies.

Solutions:

1. **Regular** Filter Maintenance: To address clogged air filters, adopt a regular filter maintenance routine. Check and change or clean filters as the manufacturer recommends, typically every one to three months. This simple step can significantly improve your system's efficiency. It is one of the easiest and most cost-effective ways to ensure your HVAC system functions optimally.

2. **Refrigerant Inspection:** If you suspect refrigerant issues, it's crucial to call a professional HVAC technician to assess and address the problem. Attempting to handle refrigerant yourself can be dangerous and may result in further issues. Professional technicians have the expertise and equipment to detect and fix refrigerant leaks, ensuring your system operates at its best. Always opt for certified technicians who follow industry best practices.

3. **Thermostat Calibration:** If your thermostat is the problem, ensure it is calibrated correctly. If necessary, replace it with a modern, programmable thermostat that can help optimize temperature settings for better energy efficiency. Thermostat issues can sometimes be resolved through simple calibration or by replacing batteries. However, if the thermostat is malfunctioning beyond your ability to fix, it's best to call a professional technician. They can assess whether it's a wiring issue or if a new thermostat is required. Modern programmable thermostats provide precise control over your HVAC system and can help you save on energy costs.

Uneven Temperature Distribution: Balancing Airflow

Causes:

1. **Obstructed Vents:** Uneven temperature distribution often occurs when vents are obstructed. Objects such as furniture, curtains, or even closed doors can disrupt airflow, leading to hot and cold spots in your living space.

2. **Ductwork Issues:** The ductwork in your home plays a significant role in delivering conditioned air to different rooms. Leaks, gaps, or inadequate insulation in your ductwork can result in temperature inconsistencies. Ductwork issues can lead to conditioned air escaping before it reaches its intended destination, causing an imbalance in your home's temperature.

Solutions:

1. **Vent Maintenance:** Ensure that all vents are unobstructed. Regularly inspect your home for objects blocking vents and rearrange furniture if necessary. You can also consider using vent deflectors to direct air where needed most. Vent deflectors can be particularly helpful in large or open spaces where air distribution can be challenging.

2. **Duct Inspection:** Have a professional inspect your ductwork for leaks, gaps, or insulation issues. Properly sealed and insulated ducts will help distribute air evenly throughout your home. Duct inspection and sealing should be part of your rou-

tine HVAC maintenance to ensure your system operates optimally. Professionals use tools like smoke pencils and infrared imaging to detect leaks that may not be visible to the naked eye. Proper sealing and insulation prevent conditioned air from escaping and help maintain uniform temperatures throughout your home.

Strange Noises: Identifying and Addressing Unusual Sounds

Causes:

1. **Rattling or Clanging:** If you hear rattling or clanging sounds from your HVAC system, it's likely due to loose or damaged parts. This can include fan blades, motors, or belts that have come loose or show signs of wear and tear.

2. **Hissing or Whistling:** Air leaks in the ductwork may cause these sounds. When there are holes or gaps in the ducts, conditioned air escapes, creating a hissing or whistling noise as it rushes through the openings.

3. **Squealing or Screeching:** These high-pitched sounds are often related to a malfunctioning blower motor or a worn-out fan belt. When the blower motor bearings wear out, it can result in a squealing or screeching noise.

Solutions:

1. **Visual Inspection:** When strange noises emanate from your HVAC system, the first step is to conduct a visual inspection. Look for loose parts like fan blades, motors, or belts. If you're comfortable doing so, you can tighten loose components. However, it's best to call a professional technician if the issue is more complex. Loose or damaged parts should not be ignored, as they can cause more significant problems if left unattended. A loose fan blade, for example, could damage the unit's interior or pose a safety hazard.

2. **Duct Sealing:** If you hear hissing or whistling sounds, especially when your HVAC system is running, it often indicates ductwork leaks. Sealing these leaks with mastic or foil tape can resolve the issue. Duct sealing is essential not only to eliminate these noises but also to prevent energy wastage. Conditioned air escaping from leaky ducts means your system has to work harder to maintain the desired temperature.

3. **Lubrication or Replacement:** When it comes to squealing or screeching

sounds, particularly those associated with the blower motor or fan belt, you may need to apply lubrication to the affected components. However, if the issue persists, it's advisable to consult a professional technician. They can determine if the blower motor needs replacement or if a worn-out fan belt is causing the noise. Attempting to lubricate or replace these parts without proper knowledge can lead to further damage. Moreover, a malfunctioning blower motor can significantly impact your system's performance and energy efficiency.

Leaking Refrigerant: Detection and Resolution

Causes:

1. **Corrosion:** Over time, refrigerant lines can corrode or develop small leaks. This corrosion can be due to factors like moisture, contaminants, or the natural wear and tear associated with the system's aging.

2. **Poor Installation:** Inadequate installation can cause refrigerant leaks, especially at connection points. When your HVAC system is not installed correctly, it can lead to issues with the refrigerant lines and connections.

Solutions:

1. **Professional Inspection:** If you suspect a refrigerant leak, it's essential to have a professional HVAC technician inspect your system. Detecting and fixing refrigerant leaks is a task best left to the experts. They can safely and accurately identify the leak's location and take the necessary steps to repair it. This typically involves repairing or replacing the damaged section of the refrigerant line and then recharging the system with the appropriate refrigerant. Attempting to address refrigerant issues alone can further damage your system and potential safety hazards.

2. **Preventative Maintenance:** Regular maintenance can help prevent refrigerant leaks in the first place. Ensure your HVAC system is installed by qualified professionals who follow industry best practices. Proper installation and routine maintenance will go a long way in ensuring that your system remains leak-free and operates efficiently. Refrigerant issues can be mitigated by choosing experienced and certified technicians who prioritize the longevity and performance of your system.

Thermostat Malfunctions: Diagnosis and Repair

Causes:

1. **Wiring Issues**: Incorrect or loose wiring can cause thermostat malfunctions. The thermostat relies on electrical connections to communicate with your HVAC system. When these connections are compromised, the thermostat may not function as intended.

2. **Battery Problems:** Some thermostats use batteries to power their operation. Depleted or dead batteries can lead to thermostat malfunctions.

3. **Outdated or Faulty Thermostats:** Older thermostats may not communicate with your HVAC system effectively, leading to inaccuracies in temperature settings and control.

Solutions:

1. **Wiring Inspection:** Carefully inspect the thermostat wiring to ensure it is correctly connected. Reestablish the connections or replace the wiring if you find loose or damaged wires. Wiring issues can sometimes be a straightforward fix, but it's best to call a technician if you're unsure or uncomfortable handling electrical connections. They can accurately diagnose and rectify the issue, ensuring your thermostat communicates effectively with your HVAC system.

2. **Battery Replacement:** If your thermostat uses batteries, replace them regularly to avoid malfunctions. The frequency of battery replacement depends on the type of batteries your thermostat uses. Still, it's generally a good practice to change them once a year or as the manufacturer recommends. New batteries ensure that your thermostat remains operational and continues to control your HVAC system accurately.

3. **Upgrading:** Consider upgrading to a modern, programmable thermostat. Modern thermostats offer precise control over your HVAC system and can be programmed to adjust temperatures according to your schedule. They also provide features like remote access and integration with smart home systems, enhancing energy efficiency and convenience. Upgrading your thermostat can be a valuable investment in your HVAC system's performance and overall comfort.

Electrical Issues: Safety Measures and Solutions

Causes:

1. **Electrical Shorts:** Electrical shorts can occur due to worn-out wires or damaged components within the HVAC system. These shorts can lead to system malfunctions, including complete shutdown.

2. **Tripped Breakers:** An overloaded electrical circuit can trip the circuit breaker, causing your HVAC system to malfunction. Overloads typically happen when multiple appliances or electrical devices operate simultaneously on the same circuit.

Solutions:

1. **Safety First:** When dealing with electrical issues, safety is paramount. Electricity can be hazardous, and if you're not comfortable working with electrical components, it's best to call a professional electrician or HVAC technician. Handling electrical problems without the necessary knowledge and safety precautions can result in personal injury or further damage to your system.

2. **Tripped Breakers:** If a circuit breaker has tripped, it indicates an electrical overload. To address this, first identify the cause of the overload. Check which appliances or devices were running when the breaker tripped. Often, this happens when too many electrical devices are connected to the same circuit. Distribute the load by connecting appliances to different circuits to prevent future overloads. After resolving the issue, reset the circuit breaker. If the breaker continues to trip, it's crucial to consult a professional electrician. They can assess your electrical system to ensure it can handle your load and make any necessary upgrades.

3. **Professional Inspection:** For electrical shorts or other complex electrical problems within your HVAC system, seeking professional assistance is crucial. Professional electricians and HVAC technicians have the training and experience to safely diagnose and repair electrical issues. They can identify the source of the problem, whether it's a damaged wire or a malfunctioning component, and take the appropriate measures to resolve it. When it comes to electrical problems, never compromise on safety, as the consequences of mishandling electricity can be severe.

DIY VS. PROFESSIONAL MAINTENANCE

In our quest to master the control of HVAC systems and save money on repairs and maintenance, one crucial decision we must make is to tackle maintenance tasks ourselves or call in a professional technician. The choice between DIY (Do-It-Yourself) and professional maintenance is a pivotal one that directly impacts the efficiency and longevity of your HVAC system.

Knowing Your Limits: When to Call in a Professional

One of the first steps in managing your HVAC system effectively is recognizing your limits. While DIY maintenance can be rewarding and cost-saving, not all tasks are suited for the average homeowner. Sometimes, calling in a professional HVAC technician is not just a wise choice; it's a necessity.

Common DIY Maintenance Tasks

Before we dive into when to seek professional help, let's first outline the common DIY maintenance tasks that homeowners can usually handle successfully:

1. Changing Air Filters

This is a simple yet crucial task that every homeowner should know. Regularly replacing air filters can significantly improve your system's efficiency and indoor air quality.

2. Cleaning and Dusting

Routine HVAC system cleaning, including vents and ducts, is a straightforward DIY task. It can prevent dust and debris from accumulating and impeding system performance.

3. Thermostat Calibration

Adjusting your thermostat settings and calibrating them according to your comfort needs is something most homeowners can do without professional assistance.

4. Inspecting for Visible Issues

A basic visual inspection of your HVAC system can help you identify problems like loose wires or leaks.

Instances Requiring Professional Intervention

While the above tasks are well within the reach of most homeowners, there are situations where professional intervention is necessary to avoid complications and ensure safety.

1. Complex Repairs

If your HVAC system is experiencing issues beyond basic maintenance, such as a malfunctioning compressor or a refrigerant leak, it's time to call a professional. Attempting complex repairs without the necessary expertise can worsen the problem and potentially endanger your system.

2. Electrical Work

Working with electricity is inherently risky. Suppose you must address electrical issues within your HVAC system, such as wiring problems. In that case, relying on a licensed technician who understands the system's electrical components is crucial.

3. Refrigerant Handling

Dealing with refrigerants requires specialized equipment and knowledge to ensure safety and compliance with environmental regulations. Handling refrigerants without proper training can lead to health hazards and environmental damage.

4. System Installation and Replacement

Installing a new HVAC system or replacing an existing one is a professional job. It involves complex calculations, sizing, ductwork, and refrigerant handling. Mistakes during installation can lead to inefficiency and expensive repairs down the road.

Knowing when to tackle maintenance tasks yourself and when to call in a professional is essential for maintaining the efficiency and longevity of your HVAC system. While DIY maintenance is valuable for routine upkeep, it's equally important to recognize your limits and seek professional help when necessary. Next, we will explore how to find a reliable HVAC technician when professional assistance is needed.

Finding a Reliable HVAC Technician: Hiring Tips

When it's time to bring in a professional HVAC technician, the quality of their work can significantly impact your system's performance and your long-term cost savings. Finding a reliable technician is crucial.

1. Check Credentials and Licensing

One of the first steps in your search for a dependable HVAC technician is to verify their credentials and licensing. This information assures you that the technician has received proper training and complies with local regulations.

- Licensing: HVAC technicians are typically required to be licensed by the state where they operate. Check the technician's license to ensure they are legally permitted to perform HVAC work in your area.

- Certifications: Look for certifications from reputable organizations like NATE (North American Technician Excellence) and ACCA (Air Conditioning Contractors of America). These certifications testify to the technician's commitment to professionalism and quality work.

2. Experience and Reputation

Experience matters in the HVAC industry. Technicians with a proven track record are more likely to handle complex issues effectively. Here are some ways to assess their experience and reputation:

- References: Ask the technician for references from past clients. Contact these references to inquire about their experiences with the technician's services.
- Online Reviews: Read Yelp, Google, and the Better Business Bureau reviews. These reviews can provide insights into the technician's reputation and customer satisfaction.
- Years in Business: Consider how long the technician or their company has been in business. Longevity can be a sign of reliability and stability.

3. Insurance and Liability Coverage

Accidents can happen during HVAC maintenance or repair work. To protect yourself and your property, ensure the technician is adequately insured. A reputable HVAC professional should have liability and workers' compensation insurance.

4. Written Estimates and Contracts

Before the technician starts any work, ask for a detailed, written estimate that outlines the scope of the job, labor and material costs, and any warranties or guarantees. This document is a clear reference for both parties and helps prevent misunderstandings.

5. Energy Efficiency Expertise

An HVAC technician with expertise in energy-efficient solutions can help you achieve your goal of saving money in the long run. They can recommend upgrading your system to more energy-efficient models or optimizing your existing system for better performance and lower energy consumption.

6. Emergency Services and Availability

Consider the technician's availability for emergency services. HVAC issues don't always occur during regular business hours. A technician who offers 24/7 emergency services can be a lifesaver when dealing with a sudden system breakdown.

7. Written Warranty

A written warranty on labor and parts can provide peace of mind. Ensure to understand the terms and conditions of the warranty before agreeing to the work.

8. Get Multiple Quotes

It's advisable to obtain quotes from several HVAC technicians before deciding. This allows you to compare prices and services to ensure you get the best value for your investment.

9. Communication Skills

Effective communication is vital for any service provider. The technician should be able to explain the issues, solutions, and costs clearly. They should also be responsive to your questions and concerns.

10. Trust Your Instincts

Lastly, trust your instincts when choosing an HVAC technician. Continuing your search is okay if something feels wrong or the technician seems unprofessional. Your comfort and peace of mind are important in this process.

Cost-Benefit Analysis: DIY vs. Professional Maintenance

Now that you understand when to call in a professional and how to find a reliable HVAC technician, let's delve into the cost-benefit analysis of DIY vs. professional maintenance. This analysis can help you make informed decisions based on your circumstances and goals.

The Cost of DIY Maintenance

DIY maintenance can be cost-effective, especially for routine tasks like changing air filters, cleaning, and simple thermostat adjustments. However, it's essential to consider the potential costs associated with DIY maintenance:

1. Time and Effort

While DIY tasks may save you money upfront, they often require significant time and effort. You'll need to research, gather materials, and perform maintenance, which can be time-consuming.

2. Risk of Errors

DIY maintenance carries the risk of errors. A small mistake can lead to larger problems, ultimately costing you more in repairs or system inefficiency.

3. Lack of Expertise

Most homeowners lack the expertise to diagnose and address complex HVAC issues. Attempting these tasks without the necessary knowledge can lead to costly mistakes.

4. Warranty Implications

Some HVAC manufacturers require professional maintenance to keep warranties valid. If you perform DIY maintenance and encounter issues, you may void your warranty, leading to expensive out-of-pocket repairs.

The Benefits of Professional Maintenance

Professional HVAC maintenance comes with several benefits that can justify the costs:

1. Expertise and Precision

Experienced technicians have the knowledge and tools to diagnose issues accurately and perform maintenance with precision, ensuring optimal system performance.

2. Time Savings

Professional technicians can complete tasks more efficiently, saving you time better spent on other activities.

3. Warranty Preservation

Professional maintenance often preserves your HVAC system's warranty. If issues arise, you're covered, reducing your long-term repair costs.

4. Preventative Care

HVAC professionals perform preventative maintenance that can identify and address potential issues before they become major problems, saving you money and headaches.

5. Enhanced Efficiency

Regular professional maintenance improves your system's efficiency, lowering energy bills and long-term savings.

Finding the Right Balance

The ideal approach is often a combination of both DIY and professional maintenance. Consider the following strategies:

1. Routine DIY Maintenance

Continue to perform routine DIY maintenance tasks like changing air filters and cleaning to keep your system in good condition.

2. Professional Annual Inspection

Schedule an annual professional inspection and maintenance visit. This ensures that your system receives expert care and that any emerging issues are addressed promptly.

3. Complex Tasks

Always rely on professionals for complex repairs or installations to avoid costly mistakes and ensure safety and compliance.

Developing a Long-Term Maintenance Plan: Balancing DIY and Professional Services

A long-term maintenance plan is essential to keep your HVAC system in peak condition and maximize cost savings. This plan should incorporate both DIY and professional services strategically and cost-effectively.

1. Establish a Regular Schedule

Consistency is key to effective HVAC maintenance. Create a schedule for both DIY and professional maintenance tasks. For example:
- Monthly: Change air filters.
- Quarterly: Perform visual inspections and clean vents.
- Annually: Schedule a professional inspection and maintenance visit.

2. Keep Detailed Records

Maintain records of all maintenance and repairs. Note the date, tasks performed, and any issues identified. This documentation is valuable for tracking your system's performance and helps technicians diagnose problems more effectively.

3. Implement Energy-Efficient Practices

Incorporate energy-efficient practices into your long-term maintenance plan. This includes adjusting thermostat settings, sealing air leaks, and considering system upgrades for improved efficiency.

4. Budget for Professional Services

Allocate a budget for professional HVAC maintenance and repairs. It's an investment that pays off by extending your system's lifespan and reducing long-term costs.

5. Stay Informed

Stay informed about industry developments and technological advancements. New HVAC technologies and energy-saving strategies can impact your long-term maintenance plan, so it's essential to adapt accordingly.

Preventive Maintenance Contracts: Pros and Cons

Some HVAC service providers offer preventive maintenance contracts, known as service agreements. These contracts are designed to provide regular maintenance and inspection services for a fixed fee. While they can be beneficial, weighing the pros and cons before committing to one is essential.

Pros of Preventive Maintenance Contracts

1. Regular Maintenance

Service agreements ensure that your HVAC system receives regular maintenance, reducing the risk of unexpected breakdowns and costly repairs.

2. Priority Service

In the event of a system issue, contract holders often receive priority service, ensuring faster response times.

3. Extended Lifespan

Regular maintenance through a contract can extend the lifespan of your HVAC system, delaying the need for costly replacements.

4. Budget Predictability

With a fixed annual fee, you can budget for HVAC maintenance without worrying about surprise expenses.

Cons of Preventive Maintenance Contracts

1. Cost

Service agreements come with an annual or monthly cost, which may not be justifiable for all homeowners, especially if you're diligent in performing DIY maintenance.

2. Quality Variation

The quality of service can vary among HVAC providers. Choosing a reputable company is crucial to ensure you receive the level of service you're paying for.

3. Locked-In Contracts

Some contracts lock you in for a specific period, which can be restrictive if circumstances change.

4. Warranty Requirements

Review your HVAC system's warranty terms. Some manufacturers require specific maintenance procedures to maintain warranty coverage. Ensure that the service contract aligns with these requirements.

PART 2
EFFICIENCY AND SAVINGS
PART 2

CHAPTER 3
Energy Efficiency and Cost Savings

As any HVAC professional or homeowner knows, energy efficiency is key for reducing costs associated with heating, ventilation, and air conditioning over the long run. Small changes like upgrading insulation, sealing air leaks, and properly maintaining systems can yield significant savings year after year without sacrificing comfort. This chapter will explore various strategies for boosting efficiency and lowering operating expenses for any HVAC system. From smart controls and equipment upgrades to home sealing techniques and available financial incentives, learning to maximize energy performance is one of the best investments you can make.

MASTERING ENERGY EFFICIENCY: STRATEGIES AND TACTICS

Improving the efficiency of HVAC systems begins by understanding exactly where and how energy is wasted. A holistic view combining building science, equipment performance, and occupant behavior change creates the strongest path toward long-term operating costs and carbon footprint reductions. While equipment upgrades help boost efficiency, low or no-cost solutions applied correctly through education and technical know-how can yield savings. This heading explores various strategies professionals and homeowners can use to take energy efficiency to the next level.

Educating Homeowners

As the experts, we must make educating homeowners a top priority. Show clients how small adjustments to things like thermostat programming, duct maintenance, and air filter changing can lower bills significantly with minimal effort. Provide reference guides and schedule follow-up sessions to reinforce efficient habits. When occupants grasp subtle efficiency opportunities, lasting behavior modification results organically. Starting conversations early also cultivates buy-in for future whole-system improvements.

Thermostat Management

Teach clients only to set back temperatures a few degrees at most when away to prevent excessive run times upon return and maintain indoor humidity control. Scheduling morning warm-up/cool-down periods prevents shock to the system and ensures interior comfort upon occupant wake-up. Demonstrating how to override programmed setpoints for parties or during off-schedule times maintains flexibility, too. Consistent thermostat use maximizes efficiency without sacrificing comfort.

Regular System Check-Ups

Walk clients through performing basic seasonal system evaluations. Recommend inspecting evaporator/condenser coils, checking capacitor performance, and lubricating motors in spring and fall. Train them to monitor filter status with a smartphone app reminder every 30 days. Come spring and fall, demonstrate checking thermostat calibrations and ductwork for leaks, too. Do-it-yourself evaluations catch minor issues before they worsen, saving big on repairs down the road.

Filter Maintenance

Dirty filters block airflow and hamper overall efficiency. Educate clients never to clean disposable filters but to replace them monthly. For cleanable filters, demonstrate the proper cleaning technique of soaking in warm water and blotting dry instead of harsh abrasives like wire brushes that damage the media. Clients should also know to double-check that the filter fits tightly after reinstalling. Proper maintenance here adds up to major utility savings in the long run.

Duct Assessment and Sealing

Show homeowners how to inspect ductwork themselves using a simple depressurization test. Explain signs of deterioration like disconnected seams, crushed conditioning, or tossed joints that should trigger professional remediation. Demonstrate safe techniques like mastic or wrapped fiberglass for minor sealing to reinforce seams versus dangerous duct tapes. Sealing major leaks slashes duct losses by huge margins, assisting overall system output.

HVAC Sizing Fundamentals

Ensure clients understand basic sizing principles related to equipment capacity, interior conditions, local climate, and building shell traits. Explain how mismatched capacities lead to premature failure or inefficient operation. Help homeowners learn signs of over- or undersizing like constant high/low runtimes or inability to reach/maintain temperature setpoints. Catching sizing issues early makes replacement decisions simpler down the road.

Zoning System Basics

Educate on multi-zone HVAC benefits like individualized heating/cooling and energy savings from isolating unoccupied areas. Demonstrate basic zones with smart vents/dampers, or separate outdoor units may be created. Guide clients on identifying candidate rooms for zoning based on usage patterns. Zoned layouts improve comfort while eliminating energy wasted on unneeded conditioning in vacant rooms.

Detecting Duct Leaks

Using a smoke pencil, demonstrate common leakage points around plenums, takeoffs, joints, and seals clients can check themselves. Explain how the pressurized smoke test method works and share a supply/return pressure test checklist for seasonal inspections. Confidently catching leaks is half the battle – showing occupants simple diagnosis techniques empowers lasting duct maintenance.

Balancing the Distribution System

Help homeowners balance their ducts by connecting a manometer and explaining how to adjust dampers until each grille reaches the proper CFM. Practice achieving +/-10 CFM on takeoffs together as a learning exercise. Balanced airflow remedies poor temperature distribution, drafts, and wasted conditioning to under-performing zones.

Conducting Load Calculations

Guide clients step-by-step through basic ACCA Manual J load calculations using freeware programs. Stress the importance of gathering home dimensions/construction data accurately. Explaining calculation principles builds trust while helping catch sizing issues independently. DIY load calculations catch mismatched equipment before breakdowns occur.

Creating a Home Energy Profile

Work with homeowners to map our home energy profile listing square footage, zoning layout, insulation levels, window treatments, appliance/lighting stats, and thermostat settings. Stress locating utility bill histories and understanding rate structures, too. Benchmark current usage against neighbors with similar homes. Profiles flag priority areas and efficiency opportunities requiring attention.

Using Energy Monitoring Tools

Introduce clients to smart home energy monitors or data loggers to track real-time and historical HVAC usage patterns. Demonstrate installation of the monitoring equipment and app interface. Analyzing trends over seasons helps recognize influences like weather, occupancy, scheduling, and system performance. Data drives customized solutions for maximum customized savings.

Infrared Imaging Basics

Loan clients an infrared camera to explore building science principles first-hand. Guide imaging exterior walls, attic hatches, windows, and doors to locate thermal bridges while the HVAC operates. Share sample images to build interpretation skills. Thermal imaging strengthens understanding of where efficiency upgrades offer the highest payback through real-world visual proof.

Education lays the foundation for sustained efficiency improvements. Minor symptomatic issues become major cost-saving opportunities by confidently demonstrating fundamental evaluation skills and empowering homeowners to monitor usage independently. Continuous guidance nurtures long-term efficiency-mindedness through every phase of ownership. Combined with strategic upgrades covered in later headings, occupant knowledge becomes the greatest asset for maximizing savings potential year after year.

UPGRADING YOUR SYSTEM FOR LONG-TERM SAVINGS

While proper maintenance keeps existing equipment running optimally, components eventually reach the end of their useful lifespan, requiring replacements or upgrades. Making well-informed decisions with efficiency top of mind maximizes returns on investment through reduced utility costs far into the future.

Replacing Aging Equipment

As furnaces, air conditioners, heat pumps, and boilers age 10-15+ years, internal components degrade, causing diminished performance. Explain signs it may be time for replacement, like inconsistent temperature control, excessively long run times, unusual noises or vibrations, and failing ignition systems. Emphasize that early replacements cost less than prolonged deferral, leading to component failures.

High-Efficiency Equipment Options

Review ENERGY STAR qualified options meeting minimum SEER, HSPF, and AFUE thresholds. Compare performance and cost ratios of air source heat pumps to gas furnaces. Suggest tankless on-demand water heaters and mini-splits where ductwork presents challenges, too. Proper sizing and a high-efficiency system provide maximum payback through lower utility bills.

Gas vs. Electric Equipment Efficiency

Explain gas furnaces reach AFUE 80-98% while electric heat pumps offer higher SEER 15-30+/HSPF 8.2-12+ depending on head/blower upgrades. Both have advantages - gas ignition provides quicker response while heat pumps efficiently produce 1.5-3 units of heat per kW using absorbed outdoor energy. Smart controls maximize performance year-round with either fuel type.

Air Conditioner Efficiency Upgrades

Review the benefits of adjusting external static pressure and adding features like two-stage compressors, ECM blowers, and variable speed components. Explain their output precision improves part-load efficiency, dehumidification, and indoor air quality. Emphasize smart HVAC controls optimized for variable capacities to deliver maximum efficiency potential.

Furnace Fan Upgrades

Compare constant torque PSC versus efficient ECM blowers. Illustrate how ECMs reduce energy use by up to 50% through supply airflow tuning. They also communicate enabling smart thermostats, HRVs, and ERVs by providing constant cfm. With a 15-20-year lifespan, fan replacements pay themselves back rapidly through long-term savings.

Proper Equipment Sizing

Performing accurate Manual J load calculations considering factors like duct location, envelope insulation levels, air changes, appliances, and people loads ensures selecting correctly sized equipment. Oversizing wastes significant energy, while under sizing leads to premature failure. Right-sizing alone delivers 10-15% savings.

Zoned Heating Systems

Review the benefits of multi-zone heating for individualized comfort control without conditioning unused rooms. Compare options like separate outdoor unit systems, smarter zone valves/dampers, or HVAC split systems incorporating internal zoning controls. Emphasize zoning maximizes efficiency in large, differently-oriented homes.

Ductless Mini-Split Heat Pumps

Introduce ductless mini-split heat pumps as an alternative to traditional central HVAC, especially when retrofitting. Demonstrate their wall-mounted indoor components condition target areas efficiently without duct losses. Encourage evaluating them for additions, basements, garages, or home heating/cooling applications.

Geothermal Heat Pump Systems

Provide an overview of geothermal or ground source heat pumps utilizing underground piping loops as an ultra-high-efficiency renewable option. Note while expensive to install initially, they achieve efficiencies 300-600% greater than air sources with a 7-10 year simple payback. Geothermal qualifies for maximum rebates and tax incentives, too.

Heat Recovery Ventilation Systems

Review the function and benefits of heat and energy recovery ventilators, bringing fresh air and transferring heat/humidity between incoming and outgoing air streams. Explain their importance alongside high-performance envelopes for optimizing indoor air quality while minimizing conditioned air losses.

Duct Insulation and Sealing

Review the significance of properly sealed and insulated ductwork for maximizing system performance. Explain common duct losses account for 15-30% of heating/cooling costs, and sealing/insulating reduces that tremendously. R-value standards avoid thermal bridging, too. Always evaluate duct improvements coinciding with equipment upgrades.

Variable Capacity HVAC Options

Compare the advantages of variable capacity equipment like variable speed air handlers, modulating furnaces, and inverters on heat pumps for improved part-load efficiency. Show they use 30-50% less energy than single-stage units through gentle temperature/humidity ramping. Advanced zoning expands their capabilities, too.

Heat Pump Water Heaters

Review electric air-source or geothermal water heaters' efficiency, comfort, and cost benefits. Note they achieve efficiencies 2-3 times greater than standard electric or gas storage tanks through using refrigerants as the heat exchange medium. Qualify for rebates and deliver $400-600 in annual savings on water and heating utility bills alone.

HVAC Control Upgrades

Review the benefits of smart WiFi thermostats, which automate schedule programming for 7-day setback operations. Explain more advanced control options integrating occupancy sensing, geofencing, and HVAC system monitoring to take efficiency even further. Communicate all controls to maximize equipment performance for maximum energy reduction and comfort control.

Building Envelope Improvements

Stress-sealing air leaks and boosting whole building insulation levels should accompany HVAC upgrades for a cohesive efficiency approach. Higher R-values block thermal bridging while sealed envelopes stop unconditioned air exchange. These foundational steps cut total energy usage by 25-40% across space conditioning and water heating.

HVAC Replacement Prioritization

Emphasize all replacement decisions require careful cost-benefit analysis. Furnish a prioritization worksheet listing replacement criteria to help clients rationally compare upgrade options. Explain variables to consider: efficiency ratings, utility rebates, existing equipment age/condition, estimated annual energy/cost savings, and payback periods. Data-driven choices maximize returns.

HVAC Replacement Project Planning

Provide a sample HVAC replacement project timeline and checklist. Include suggested spring/summer targeted replacement windows, lead times for ordering high-efficiency units, scheduling ductwork/control retrofits accordingly, and lining up required permits in advance. Proper planning ensures smooth, cost-effective installations capturing peak cooling season.

Evaluating Current Controls

Properly evaluating a home's existing controls is the first step to determining appropriate upgrade recommendations. Start by inspecting the current thermostat for any issues inhibiting efficient operation. Ensure it is firmly mounted and has fresh batteries, if applicable. View the programming and note how easy the schedule is to understand and change. See if temporary overrides are being overused, indicating the current programming may not be optimal. Examine where the thermostat is located and ensure the area receives average temperatures for the house. Any red flags, such as difficulty programming or a location in direct sun, should be documented.

Record runtimes over a day using equipment monitoring devices. Note how long the system runs during each cycle and compare it to expected runtimes. Unusually short or long runtimes could point to improper maintenance, sizing concerns, or suboptimal thermostat placement. Evaluate satisfaction with the current temperature stability and humidity control. Oscillating temperatures or comfort issues highlight inefficiencies to address. Overall, thoroughly examining existing controls sets the stage for determining what smart upgrades are needed.

Programmable vs. Smart Thermostats

Programmable thermostats offer scheduling capabilities but require manual programming with no remote access or automatic features. Budget-friendly programs are better than nothing, but engagement and continuous optimization are limited compared to smart options. Programable also lack sensors that detect changes like opening windows or doors and suspend conditioning accordingly. A programmable thermostat may be a temporary solution for clients seeking convenience and energy savings typical of smart models, but upgrading is recommended.

Smart, Wi-Fi-enabled thermostats take automation and optimization to the next level. Geofencing tracks homeowners' phones to heat and cool only when occupants are home. Automatic scheduling analyzes usage patterns to generate setpoints, maximizing comfort and efficiency. Sensors allow suspending HVAC if windows are open, preventing wasted energy. Integrations with voice assistants and smart home platforms provide ultimate remote access from anywhere via phone or tablet. Advanced analytics even send maintenance alerts if performance declines are detected. Few options exist to reduce energy bills as effortlessly as a modern smart thermostat.

Programming Fundamentals

Consistency is key to maximizing savings when programming any basic or smart thermostat. Walk through establishing a seven-day schedule, showing how to set individual nighttime setbacks for weekday vs weekend programming if desired. As a rule, modest setbacks of just 2-3°F for heating and 2-3°F higher for cooling are enough to substantially reduce run times without compromising comfort. It's also important to enable temporary hold features, demonstrating how these can comfortably override the schedule for parties or off-schedule wake/sleep periods without disrupting the regular program.

Programming schedules should also factor in climate and system variables. In very hot or cold climates, slightly smaller setbacks of 1-2°F may suffice to start. Likewise, set morning pre-cool or pre-heat recovery periods before wake-up to prevent shocking occupants. For multifamily homes, advise adjusting staged heating/cooling start times based on floor or wing locations. Always have homeowners try out draft schedules for at least two weeks before finalizing to ensure temperatures remain acceptable throughout occupancy hours. With attentive programming education, clients leave ready to maximize efficiency from day one.

Achieving Optimal Comfort

While programming aims to reduce energy usage, maintaining comfort should never be compromised. Use any initial discomfort issues to reinforce factors influencing interior conditions. Review thermostat placement fundamentals, emphasizing that a statistic reads the air in its immediate area, not the whole house average. Inspect for proper airflow and system sizing match, too. Calibration differences of only half a degree can impact the perceived comfort of setpoints.

Train homeowners to try adjusting setpoints gradually to pinpoint personal comfort zones before assuming an equipment or installation issue exists. Keeping a comfort journal noting indoor/outdoor conditions, activities, and clothing levels helps identify patterns, too. Consider occupancy, schedule, or lifestyle changes straining the system's capacity. More extensive tests or upgrades should be considered only after eliminating adjustable causes through education. The goal is an efficiency-focused setup that feels uniformly comfortable for all household members.

Customizing Setback Strategies

Rather than a one-size-fits-all approach, work with clients to devise a personalized schedule mirroring their daily routine as closely as possible. Use scaling questions to determine busy versus lazy days, variable wake-up times, work-from-home patterns, or other oddities. Map out a sample week representing average schedules to emulate.

For example, a family with school-aged children would have different morning and afternoon programming than DINK professionals. Alter proposed setpoints based on client feedback, too. Always emphasize testing drafts for an extended period. Custom schedules foster optimal comfort while satisfying unique lifestyle needs. Fine-tuning improves over time as occupancy and seasons change.

Geo-Locating Smart Thermostats

Explain to customers how geo-fencing works, using Bluetooth to link the thermostat to their phones. When all phones linked to the system leave the vicinity of the home, the system knows it can raise cooling setpoints for partial hours of use. Conversely, as someone approaches the house, the system can begin to pre-cool. Demonstrate enrolling phones in the thermostat's app and checking the home/away detection range shown on a map. Geo-fencing provides automated savings without disrupting routines. Ensure customers understand occasional false positives based on signal strength.

Set up the thermostat and test the home/away feature, having customers simulate leaving to ensure proper recognition. Also, test any pre-conditioning settings to understand timing. Discuss optimum preheat/precool durations balancing comfort versus efficiency. Reinforce that the thermostat will fine-tune timing over time based on usage data. With automation handling setpoint changes, clients benefit from maximum savings with zero engagement.

Programming for Vacations

When traveling, the key is preventing wide temperature swings that could damage pipes while saving energy. Programming a narrow 2-4° set point range simultaneously accomplishes both. Show customers how to use the app to set temporary holds before/after trips without disrupting the regular schedule. For longer vacations, switching the system to "home" mode maintains routine operation using overnight setbacks rather than full shut-offs, which are less efficient. Always stress-testing vacation holds beforehand to confirm comfort is preserved upon return.

Optimizing Thermostat Placement

To ensure proper placement, review factors affecting thermostat readings. Demonstrate how drafts, direct sunlight, or heat-producing appliances like televisions and lights can skew the temperature detected. Suggest mounting locations at least 3 feet from these sources at waist level on an interior wall. Test different positions with a thermometer to find the average room temp. Photograph and note the location chosen in the homeowners' records. Placement verification prevents discomfort from improper reads.

Maximizing Energy Savings Features

Take time explaining each specialized sensor and scenario programmed into smart thermostats. Provide examples like suspending conditioning if doors remain open over five minutes. Discuss options for integrating door sensors that trigger hold periods for recurring openings like garage doors. Also, cover occupancy-linked temperature adjustments dropping to an "away" setpoint 10 minutes after the home leaves. Apps now inform users of all activities to develop an understanding of each subtle efficiency.

HOME INSULATION AND SEALING TECHNIQUES

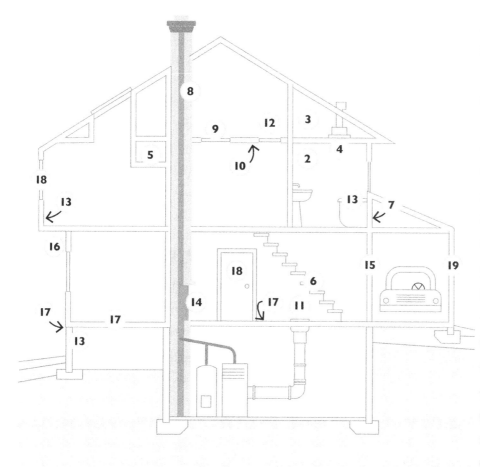

Air Sealing Trouble Spots

1. Air Barrier and Thermal Barrier Allignment
2. Attic Air Sealing
3. Attic Kneewalls
4. Shaft for Piping or Ducts
5. Dropped Celling/Soffit
6. Staircase Framing at Exterior Wall
7. Porch Roof
8. Flue or Chimney Shaft
9. Attic Access
10. Recessed Lighting
11. Ducts
12. Whole-House Fan
13. Exterior Wall Penetrations
14. Fireplace Wall
15. Garage/Living Space Walls
16. Cantilevered Floor
17. Rim Joists, Sill Plate, Foundation, Floor
18. Windows & Doors
19. Common Walls Between Attached Dwelling Units

Attic Insulation

Attics are priority areas as the largest heat gain/loss source and are prone to air leakage. Check existing R-values and depth, noting developmental standards requiring R-38+. Loosen and fluff batts if compressed to restore full coverage. Top off with loose-fill if needed to reach the code. Calculate required amounts based on blueprints for uninsulated attics and install them to proper grade in multiple layers for easy DIY installation.

Attic hatches must also be sealed tightly. Install quality gaskets and weatherstripping if inadequate, caulking seams for an air-tight barrier. Inspect for bypasses and plumb exterior vents to promote convection current. Infrared cameras find thermal bridges to further address. A well-insulated, airtight attic significantly improves home efficiency.

Wall Insulation

Walls are another major heat conductor, likely needing upgrades if a home was constructed before 1990. Inspect electrical sockets, plumbing, and attic/basement access points for air sealing. Blown-in cellulose provides an effective R-value while filling cracks and crevices. For finished walls, dense-pack cellulose forces air out during installation. Proper wall insulation prevents drafts and thermal bridging.

Floor Insulation

Crawl spaces and slab-on-grade foundations especially need attention. Depressurization testing reveals specific leaks to seal with spray foam before insulating. Roll or blown-in options raise floor R-values, slowing heat transfer. A rigid foam board reinforces the barrier against ground temperatures. A sealed, insulated floor eliminates drafts and massive energy losses.

Potential Energy Savings from Sealing and Insulating

Duct sealing alone can slash heating and cooling costs by 8-18% on average. Combining ductwork and air sealing projects captures even larger utility reductions, preventing over 25-40% of conditioned air from escaping prematurely. When attics, walls, and floors receive proper insulation, the home's performance improves dramatically. In many cases, total energy usage drops by 30-50%, representing dramatic long-term savings.

Tax Credits for Energy-Efficient Home Improvements

The Residential Energy Efficient Property Credit applies to homeowners completing qualifying insulation work. Contractors file Form 8908 reporting project costs, allowing owners to claim 10-30% of expenses on their returns. Credits range from $50 to $300 per improvement, with no overall limit. Comprehensive air sealing or insulation projects typically qualify based on presumed energy reductions.

Identifying Air Leaks

Handheld blower door tests precisely find leak locations through delicate air pressure imbalances. Infrared cameras complement by visualizing thermal bridging. Beginning attic inspection reveals bypasses around top plates, plumbing, and wiring holes. Crawlspaces next uncover foundation cracks while kneeling. Point out draft solutions like caulk, spray foam, or rope caulk.

Sealing air leaks prevents conditioned air from escaping and outside air from infiltrating. Locating major sources requires integrated diagnostic methods. Blower doors pressurize homes, using smoke or thermal imaging to visualize exfiltration points. Infrared cameras highlight thermal bridges while HVAC operates. Inspecting the outside and inside structural voids thoroughly finds every leakage avenue. Proper air sealing delivers immense energy returns.

Sealing Methods

Discuss various sealants benefits. For most applications, caulking and spray foams provide durable, airtight barriers. Demonstrate caulking techniques around windows, sill plates, and wiring holes. Spray foams require certification but form impermeable seals around plumbing, ducts, or irregular gaps. Gaps over 1/4" wide call for backer rods. Low-pressure-expanding door/window kits seal comprehensively. Matching seals to projects simplify DIY installation.

Duct Sealing Techniques

Inspecting ductwork for leaks using pressure tests, smoke, or thermal imaging. Address sections near unconditioned areas first. Common problem spots include plenum connections, takeoffs at registers, and damaged flex ducts. Mastics excel at sealing plenum seams, while minimally expanding sprays work well for other sheet metal joints. Plenums can also use mesh/mastic combos. Teach gauging proper sealant thickness.

Balancing the Distribution System

Balancing airflow optimizes comfort while preventing wasted conditioning of under-served rooms. Using diagnostic tools like duct blasters, manometers, or flow hoods, demonstrate pressure/flow tests on supply takes. Show adjusting dampers for uniform +/-10 CFM delivery. Involving clients in the balancing process improves inspection/upkeep skills long-term. Correcting imbalances maximizes system efficiency.

HVAC Equipment Rebates

Major electric and gas utilities allocate sizable rebate budgets each year towards encouraging customers to upgrade aging HVAC systems with high-efficiency alternatives. Rebates typically range from $150-500 depending on the baseline efficiency of existing equipment and end-use fuel type.

For example, replacing a standard gas furnace from the 1990s with a 95% AFUE condensing model would net $300-400 back from most utility providers. Larger rebates of $500-1000 are also commonly offered when simultaneously upgrading central air conditioners or heat pumps to SEER 16+ and HSPF 8.5+ qualifying units.

Always check the sponsoring utility's rebate webpage for incentive tiers and application procedures. Tiers increasing with higher AFUE, SEER, and HSPF ensure the most energy-efficient upgrades offer the deepest discounts. Combining HVAC replacements with additional envelope upgrades may provide additional custom rebates, too.

Smart Thermostat Rebates

Transitioning to a smart, WiFi-enabled thermostat offers potentially huge energy savings through automated setbacks and other efficiency features. Many utilities provide $50-100 statement credits to boost adoption for installing qualifying smart models.

Thermostats meeting minimum specifications for networking/programming capabilities from manufacturers like Nest, Ecobee, or Honeywell are pre-approved for instant rebates. Contractors can register the new product serial number online or via paperwork submission.

Rebates aim to cover a portion of smart thermostat hardware costs and entice customers with ease of automation. When paired with a professional installation ensuring correct placement, calibration, and programming, the potential energy reductions of up to 30% annually far outweigh rebate amounts.

Home Performance Rebates

State agencies administer deeper home retrofit rebates up to $2000-3000 to encourage energy-conscious renovations. Comprehensive projects may include air sealing, attic/wall insulation, high-efficiency HVAC, water heating, and appliances.

Customers submit a BPI-certified energy audit outlining a custom scope of work and projected cost savings. Upon completion, a post-retrofit audit verifies work quality before issuing the rebate check. More rigorous quality assurance prevents wasted funds.

Larger whole-home rebates demand a higher level of air tightness to be achieved. For example, final tests must confirm a maximum ACH50 number under 5.0 versus a simpler duct sealing project alone. Proper project scoping meets both efficiency standards and rebate qualifications.

Heat Pump Rebates

Heat pumps convert every $1 spent on electricity into approximately $3 worth of heating and cooling, making them an extremely cost-effective HVAC option. Many state energy offices provide equipment and installation rebates up to $1000-2500 to incentivize the switch from standard systems.

Rebates target minimum heat pump efficiency levels of HSPF 9.0 for air-source units and COP 3.0 for geothermal. Contractors collect serial numbers and spec sheets before filing them online or mailing them in. Applications require verifying the home has no other primary heating source like natural gas.

Ductless mini-split heat pump systems also receive generous rebates of $500-1000 per indoor unit due to their zoning benefits. Qualified installations significantly reduce upfront costs, enhancing the units' payback period attractiveness for customers.

Income-Qualified Programs

Lower-income families may access deeper rebates and even free equipment replacement for heating, cooling, insulation, or air sealing through state-run energy assistance initiatives. Eligibility is based on household size and income falling below 200% of poverty guidelines.

Approved contractors can refer income-qualified clients to energy office partners administering the program. Extensive paperwork documents proof of residency and finances. Upon approval, customers receive a consulate scheduling the free retrofit work. No upfront costs are required.

These programs use federal and state tax funds to reduce participants' annual energy burden by fully subsidizing efficiency upgrades. Automatic bill savings put more dollars in families' pockets each month.

CHAPTER 4

Common HVAC Problems and Solutions

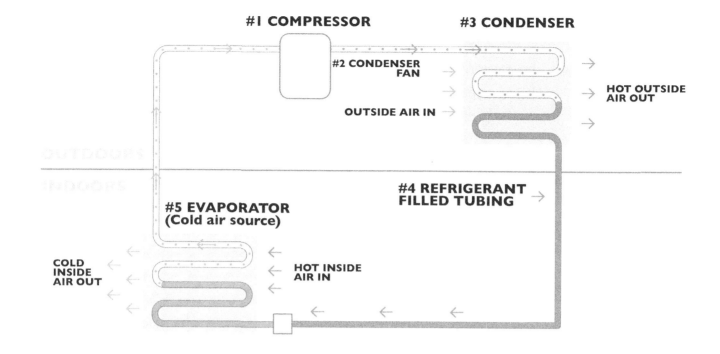

Understanding the HVAC System Diagram: Key Components

Before you embark on the troubleshooting journey, you must fully grasp your HVAC system's architecture. Just as an engineer needs to understand the blueprints of a building before fixing a structural issue, you need to comprehend the fundamental components of your HVAC system. Here, we'll delve into the key elements that make your system tick.

The Refrigeration Cycle

The refrigeration cycle is at the heart of every air conditioning and refrigeration system. This cycle is the core mechanism responsible for cooling and dehumidifying the air. Understanding it is pivotal in troubleshooting insufficient cooling or odd temperature fluctuations.

1. Compressor: The compressor is the workhorse of your HVAC system. It pressurizes and circulates the refrigerant, allowing it to absorb and release heat. Issues with the compressor often result in poor cooling performance.

2. Condenser Coil: The condenser coil releases the heat absorbed by the refrigerant. It can't do its job efficiently if it's dirty or obstructed.

3. Evaporator Coil: This coil is where the magic happens. It's where the refrigerant absorbs heat from the indoor air, causing the air to cool down.

4. Expansion Valve: The expansion valve regulates refrigerant flow into the evaporator coil, controlling the cooling process. Problems here can lead to temperature imbalances.

Air Handling Unit (AHU)

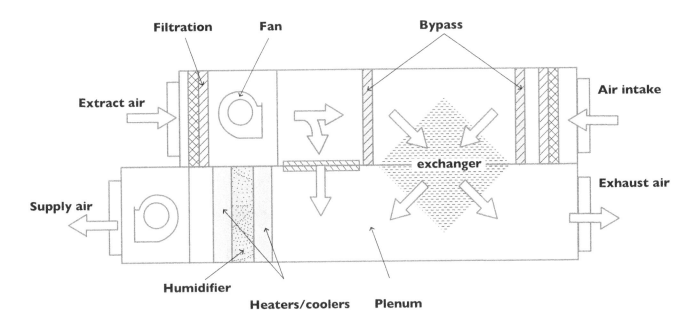

Your HVAC system isn't just about cooling or heating; it's also about air circulation. The Air Handling Unit (AHU) plays a critical role in ensuring that the conditioned air reaches every nook and cranny of your home.

1. Blower Motor: This component propels the air through the system. Problems with the blower motor can result in weak airflow and uneven temperature distribution.

2. Air Filters: Clean air filters are essential for maintaining good indoor air quality and system efficiency. Clogged filters can lead to poor performance and increased energy consumption.

3. Ductwork: Your ductwork acts as a highway for the conditioned air to travel through your home. Leaks or blockages in the ducts can cause significant issues, from energy wastage to uneven heating or cooling.

Reading Error Codes: Deciphering Diagnostic Messages

Modern HVAC systems have intelligent control systems that diagnose and report issues using error codes. These codes are your system's way of talking to you and knowing how to interpret them can save you time and money.

Common Error Codes

- **E1**: This error often relates to the condenser coil temperature sensor. If you see this code, it might be time to check for obstructions or a malfunctioning sensor.
- **F1**: The F1 code usually points to a communication issue between the various system components. It's like your system saying, "We need to talk."
- **H2**: This error indicates a problem with the temperature sensor. It's essential to address this quickly to prevent further issues.

Deciphering the Codes

Different HVAC manufacturers have their coding systems, so it's crucial to consult your system's manual for specifics. However, some general principles apply:

- **Two-Part Codes:** These are typically composed of a letter and a number. The letter indicates the type of problem (E for electrical, F for communication, H for heat, etc.), while the number specifies the specific issue.
- **Flashing LEDs:** In many systems, error codes are displayed through flashing LEDs on the control panel. The number of flashes and the timing can tell you which error is occurring.
- **Resetting Codes:** A simple power cycle can sometimes reset the error code. However, digging deeper into the problem is essential if the issue persists.

Measuring Airflow: Tools and Techniques

Inadequate airflow is a common issue in HVAC systems, and it can lead to reduced cooling or heating performance, uneven temperatures, and increased energy consumption. To troubleshoot airflow problems, you must understand how to measure them and identify potential obstructions or restrictions.

Tools for Measuring Airflow

- Anemometer: This device measures the velocity of the air. It's useful for assessing the speed of the air coming out of your vents.
- Pitot Tube: A Pitot tube is a more precise instrument for measuring air velocity and pressure. It's often used in professional HVAC diagnostics.

- Differential Pressure Manometer: This tool helps you measure pressure differences between two points in the system, which can be crucial for diagnosing airflow problems.

Measuring Airflow

1. **Inspect the Air Filter:** Start by checking the condition of your air filter. A clogged filter can significantly restrict airflow.
2. **Use an Anemometer:** Place the anemometer in front of a vent and measure the airflow. Compare this measurement to your system's specifications. If it's lower than expected, you might have an airflow problem.
3. **Check Ducts for Blockages:** Inspect your ductwork for any obstructions or damage. Sometimes, a piece of insulation or debris can block the airflow.
4. **Balance Dampers:** If you have adjustable dampers in your ductwork, make sure they are properly balanced to ensure even airflow throughout your home.

Thermographic Imaging: Detecting Heat Loss and Cold Spots

Thermographic imaging, often called thermal imaging, is a powerful diagnostic tool that can help you identify temperature variations in your HVAC system and home. By detecting heat loss or cold spots, you can pinpoint issues such as poor insulation, air leaks, or problems with your heating and cooling system.

How Thermographic Imaging Works

Thermographic cameras capture the infrared radiation emitted by objects. Warmer objects emit more radiation, while colder objects emit less. This information is displayed as a thermal image, with varying colors representing different temperatures.

Common Applications

- Detecting Insulation Issues: You can use thermographic imaging to identify areas in your home with poor insulation where warm air is escaping or cold air is infiltrating.
- Finding Air Leaks: Thermal imaging is excellent for locating gaps or cracks in your home's structure where air can leak in or out.
- Troubleshooting HVAC Systems: Thermal imaging can reveal problems with your HVAC system, such as blocked ducts or malfunctioning components.

Using Thermographic Imaging

1. Invest in a Thermal Camera: You can purchase a thermal camera or hire a professional to perform a thermographic scan of your home.
2. Inspect Your Home: Carefully examine your home's interior and exterior, focusing on areas prone to heat loss or gain, such as windows, doors, walls, and the ceiling.
3. Check Your HVAC System: Run your HVAC system and use the thermal camera to inspect components like the ductwork, vents, and heat exchangers. Temperature anomalies can indicate issues that need attention.
4. Take Action: Once you've identified problem areas, you can address them, whether adding insulation, sealing gaps, or scheduling HVAC maintenance or repairs.

Manometer Usage: Checking Gas Pressure

Gas-powered HVAC systems, including furnaces and some water heaters, rely on specific gas pressures to operate efficiently and safely. If the gas pressure is too high or too low, it can lead to performance issues, safety concerns, and even equipment damage. A manometer is the tool of choice for checking gas pressure.

U-TUBE MANOMETER

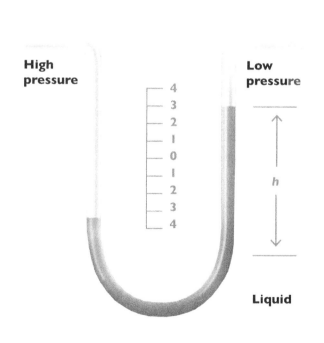

BOURDON-TUBE GAUGE

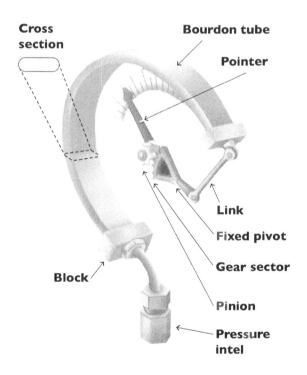

Types of Manometers

1. U-tube Manometer: This is a traditional and straightforward manometer. It consists of a U-shaped tube filled with a liquid (usually mercury or colored water). The difference in height between the liquid levels in the two legs of the U-tube indicates the pressure.

2. Digital Manometer: Digital manometers provide accurate pressure readings and are easier to read. They are especially useful for checking gas pressure in HVAC systems.

Checking Gas Pressure

1. Safety First: Before performing any work on a gas system, turn off the gas supply and follow all safety precautions.

2. Locate the Gas Valve: Find the gas valve on your HVAC system. It should be labeled and easy to access.

3. Connect the Manometer: Attach the manometer to the gas valve, ensuring a secure and airtight connection.

4. Check the Pressure: Open the gas valve and observe the manometer. It should display the gas pressure in inches of water column (in. WC). Refer to your system's specifications to ensure the pressure is within the recommended range.

5. Adjust if Necessary: If the pressure is outside the acceptable range, you may need to adjust it using the gas valve's regulator. Consult your system's manual or a professional technician for guidance.

Remember that working with gas systems can be dangerous. If you are not confident in your ability to check or adjust gas pressure, it's advisable to seek the assistance of a qualified HVAC technician.

Clogged Air Filters: Replacement and Cleaning

Air filters are the unsung heroes of your HVAC system. They help to maintain the quality of the air you breathe and ensure the smooth operation of the equipment. Over time, these filters can become clogged with dust, debris, and particles, leading to reduced airflow, decreased system efficiency, and higher energy bills.

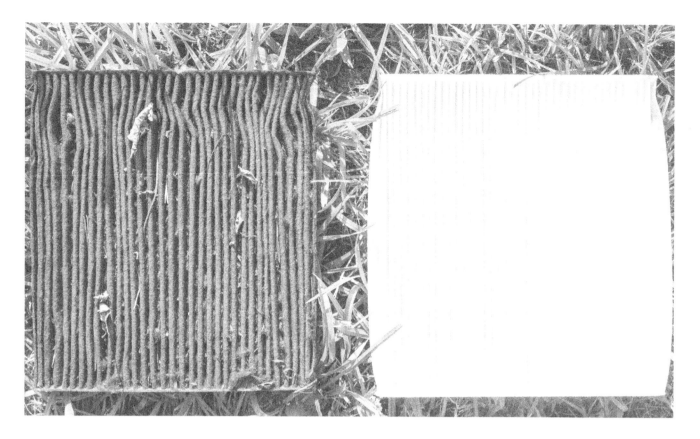

DIY Solution:

1. Turn off the System: Safety first! Before you begin, switch off your HVAC system to avoid any accidents.

2. Locate the Filter: Most filters are near the blower or in a slot within the return air duct. Check your HVAC system's manual if you're unsure.

3. Remove the Filter: Carefully slide out the old filter. If it's excessively dirty, it's best to dispose of it.

4. Inspect the Filter: Examine the filter for damage. If it's in good condition, you can try cleaning it. If it's worn or damaged, replace it with a new one.

5. Cleaning the Filter: If you clean it, vacuum or rinse it with warm water. Allow it to dry completely before reinstalling.

6. Reinstall the Filter: Carefully slide the cleaned or new filter back into its slot, ensuring it's in the right direction (usually, an arrow indicates the airflow direction).

7. Switch the System On. Turn your HVAC system back on and enjoy improved airflow and efficiency.

Frozen Evaporator Coils: Causes and Solutions

Frozen evaporator coils can indicate various problems, such as low refrigerant levels, restricted airflow, or a malfunctioning blower. When the coils freeze, your system's ability to cool your home is severely compromised.

DIY Solution:

1. Turn off the System: Safety is paramount. Always switch off your HVAC system before attempting any repairs.

2. Locate the Coils: The evaporator coils are typically found in the indoor unit near the air handler. Consult your system's manual if needed.

3. Identify the Cause: Determine why the coils have frozen. Common causes include restricted airflow, low refrigerant, or a malfunctioning blower.

4. Thaw the Coils: Allow the coils to thaw naturally or use a gentle heat source, such as a hairdryer, on a low setting. Do not use excessive heat to avoid damaging the coils.

5. Address the Underlying Issue: Once the coils are thawed, fixing the root cause is crucial. If it's a dirty filter, replace or clean it. If it's low refrigerant or a blower issue, it's best to consult a professional.

Condensate Drain Blockage: Clearing the Line

Condensate drain blockages can occur due to algae, debris, or mold accumulation in the drain line. When the drain is clogged, water can go back into the system, leading to water damage and potential system malfunction.

DIY Solution:

1. Turn off the System: Switch off your HVAC system to ensure safety.

2. Locate the Drain Line: The drain line usually runs from the indoor unit to the outside of your home. Could you find it and inspect it for blockages?

3. Clear the Blockage: There are several methods to clear the blockage. You can use a wet/dry vacuum, a pipe cleaner, or a mixture of water and vinegar. Be gentle to avoid damaging the drain line.

4. Flush with Water: After clearing the blockage, flush the drain line with water and vinegar to prevent future build-up.

5. Check for Proper Drainage: Turn your HVAC system on once you've cleared the line and ensure water drains properly.

Pilot Light Problems: Relighting and Safety Measures

A pilot light is essential for ignition for gas-powered HVAC systems. If the pilot light goes out, your system won't work. This can happen for various reasons, including drafts, thermocouple issues, or gas supply problems.

DIY Solution:

1. Turn off the Gas: Safety is paramount when dealing with gas. Turn off the gas supply to your HVAC system.
2. Locate the Pilot Light: The pilot light is typically found near the gas control valve. Consult your system's manual for guidance.
3. Follow the Manufacturer's Instructions: Different systems have specific instructions for relighting the pilot light. Follow these carefully.
4. Light the Pilot Light: Using a long lighter or matchstick, carefully light the pilot light according to the manufacturer's instructions.
5. Observe and Test: After relighting, observe the flame to ensure it's stable and blue. If it's flickering or a different color, it could indicate a problem. In such cases, it's best to contact a professional.

Non-Functioning Thermostat: Calibration and Battery Replacement

A malfunctioning thermostat can lead to temperature inconsistencies and discomfort in your home. Sometimes, the problem is as simple as a dead battery or a calibration issue.

DIY Solution:

1. Replace Batteries: Many thermostats are battery-powered. If your thermostat is unresponsive, the first step is to replace the batteries. Follow the manufacturer's guidelines for battery replacement.
2. Calibrate the Thermostat: If replacing the batteries doesn't solve the issue, your thermostat may need calibration. Consult your thermostat's manual for instructions on how to calibrate it.

3. Check the Wiring: Ensure that the thermostat wires are properly connected. If there's a loose or disconnected wire, reattach it securely.

4. Test the Thermostat: After replacing the batteries, calibrating, or checking the wiring, test your thermostat to ensure it's functioning correctly.

Air Duct Leaks: Identifying and Sealing

Leaky air ducts can significantly lose conditioned air, reducing efficiency and increasing energy bills. Identifying and sealing these leaks can substantially impact your HVAC system's performance.

DIY Solution:

1. Turn off the System: Before working on your air ducts, turn off your HVAC system.

2. Locate the Leaks: Inspect your ducts for visible gaps, holes, or disconnected sections. Common problem areas include joints and seams.

3. Sealing with Duct Tape: You can seal small gaps and holes using HVAC duct tape. Ensure the tape is designed for this purpose.

4. Mastic Sealant for Larger Leaks: Use mastic sealant for larger gaps and seams. Apply it generously and use a brush or hand to ensure a secure seal.

5. Inspect and Test: After sealing the leaks, inspect your work to ensure no gaps. Turn your system back on and check for improved airflow and efficiency.

By addressing these common issues with DIY fixes, you save money and better understand your HVAC system. However, it's essential to know your limits. If a problem goes beyond your DIY capabilities or you are unsure about any aspect of the repair, it's best to consult a professional HVAC technician. Your safety and the long-term health of your HVAC system should always be the top priority.

Becoming a master of HVAC control should involve understanding how to maintain and optimize your system and recognizing when a situation calls for expert intervention. While DIY maintenance is essential, there are times when the complexities and potential risks associated with HVAC systems demand the skills and knowledge of a professional technician.

The Importance of Timely Intervention

The adage "prevention is better than cure" couldn't be truer in the context of HVAC systems. Early detection of problems can save you from costly repairs and prevent dangerous situations. Timely intervention safeguards your investment and ensures your comfort and safety.

Gas Odors and Leaks: Immediate Action Required

When it comes to gas-related issues, there's no room for hesitation. If you ever detect the distinct odor of gas in your home or around your HVAC unit, do not delay; it's an emergency. Gas leaks are not only a threat to your property but also to your life. Natural gas is highly flammable and inhaling it can be fatal. In such cases, evacuate your home immediately, leaving the doors open to allow the gas to disperse, and contact your gas company and HVAC technician for assistance.

Your HVAC technician will inspect the unit to identify the source of the leak and repair it promptly. They will also check for any damage to the gas lines, ensuring your system is safe to operate. Gas leaks are not something to be handled by amateurs, as they require specialized equipment and expertise.

Electrical Issues: Flickering Lights and Circuit Breaker Trips

Electrical problems can be subtle yet significant. If you notice your lights flickering or your circuit breaker frequently tripping when your HVAC system is running, it's a sign of an underlying issue. These problems may indicate issues with the electrical components of your HVAC system, such as the wiring, capacitors, or compressor.

Handling electrical components without the proper training and tools can be extremely dangerous. There's a risk of electrical shock, fire hazards, or even further damage to your HVAC system. Calling a professional technician is the safest and most sensible course of action. They can identify and rectify electrical problems while ensuring your system operates safely.

Refrigerant Leaks: Environmental and Efficiency Concerns

Refrigerant is a vital component of your air conditioning system. It's responsible for cooling the air that circulates through your home. If you suspect a refrigerant leak, you might notice a decrease in cooling efficiency, the amount of cool air produced, or even ice forming on the evaporator coils.

Refrigerant leaks are problematic for two main reasons. Firstly, they harm the environment. Many refrigerants, such as older chlorofluorocarbon (CFC) and hydrochlorofluorocarbon (HCFC) refrigerants, are ozone-depleting substances. Even newer refrigerants can contribute to global warming when released into the atmosphere. Secondly, a refrigerant leak can significantly reduce the efficiency of your HVAC system, resulting in higher energy bills.

A professional HVAC technician can identify the leak's location, repair it, and recharge the system with the correct amount of refrigerant. This ensures your system operates efficiently while minimizing its environmental impact.

Persistent Strange Noises: Bearing Failure or Fan Issues

Unusual sounds emanating from your HVAC system should not be ignored. Common noises include rattling, squealing, grinding, or banging. These sounds can indicate issues like bearing failure, worn-out components, or problems with the fan motor.

If you encounter persistent strange noises, it's an indication that a part of your HVAC system is under duress and might fail if left unattended. Diagnosing and repairing these issues without proper knowledge and tools can lead to further damage. A professional HVAC technician can pinpoint the noise source and address the underlying problem, preventing further complications.

Irreparable Heat Exchanger Damage: Safety Hazards

Heat exchangers are a crucial component of your furnace, transferring heat from the combustion chamber to the air circulating through your home. Over time, heat exchangers can develop cracks or other forms of damage. These cracks can allow toxic carbon monoxide gas to escape your home's air supply, posing a severe health risk.

Detecting heat exchanger damage is beyond the scope of most homeowners. A visual inspection may not reveal the extent of the issue. An HVAC technician, however, can use specialized tools like combustion analyzers to detect carbon monoxide levels and identify any damage to the heat exchanger. If damage is detected, replacing the heat exchanger is imperative to maintain your system's and your family's safety.

PREVENTING FUTURE PROBLEMS - PROACTIVE MEASURES

The Value of Regular Inspections

The Value of Regular Inspections Imagine a scenario where you diligently care for your car, taking it for regular check-ups and addressing minor issues before they become major problems. Your HVAC system is no different. Like your car, regular inspections require identifying potential problems before escalating. Regular inspections offer several advantages: early problem detection allows you to catch issues early, preventing small glitches from becoming costly and causing major malfunctions. A well-maintained system typically has a longer lifespan. By addressing problems promptly, you're likely to avoid costly replacements. Inspections can help keep your system running at peak efficiency, saving you money on energy bills. Knowing your HVAC system is in top shape ensures it will continue to function optimally.

Implementing a Preventive Maintenance Schedule

The foundation for preventing future HVAC problems is a well-structured preventive maintenance schedule. Much like a doctor's appointment, a schedule ensures you regularly address your HVAC system's needs. Set a regular interval: depending on your system and usage, schedule inspections at appropriate intervals. Typically, twice a year, before the start of summer and winter, is a good benchmark. Maintain a record of each inspection, noting any issues, repairs, or parts replacements. This record will help in tracking your system's health over time.

While there are DIY maintenance tasks, it's advisable to have a professional HVAC technician perform a comprehensive inspection at least once a year. They have the experience and tools to identify issues you might overlook. Pay attention to key components such as the condenser coils, evaporator coils, refrigerant levels, and electrical connections. Regularly replace air filters, vital for maintaining good indoor air quality and system efficiency. Ensure that moving parts are well-lubricated.

This minimizes wear and tear and helps your system operate smoothly. Verify that your thermostat is calibrated correctly to avoid erratic temperature fluctuations. By adhering to a preventive maintenance schedule, you're taking the first step in preventing future problems and maximizing the efficiency of your HVAC system.

Keeping the Outdoor Unit Clean: Debris Removal

The outdoor unit of your HVAC system is exposed to the elements year-round. Leaves, dirt, and debris can accumulate in and around it, leading to various problems. Regularly cleaning the outdoor unit is a proactive measure to ensure the continued operation of your system. Trim any bushes, shrubs, or trees that might infringe on the unit. Make sure there's a clear space around it for proper airflow. Periodically inspect the outdoor unit for leaves, grass, and other debris and gently clean these away using a soft brush or a hose.

In colder climates, consider covering the outdoor unit in winter to prevent snow and ice from accumulating. Be sure to remove the cover before the cooling season begins. While cleaning, inspect the unit for any signs of physical damage, such as bent fins or dents which can affect efficiency and should be addressed promptly. By keeping the outdoor unit clean, you're not only maintaining the efficiency of your system but also preventing potential damage from accumulating debris and harsh weather conditions.

Upgrading Insulation: Energy Efficiency and Comfort

A well-insulated home is essential for energy efficiency and comfort. Poor insulation can cause your HVAC system to work harder, increasing energy consumption and costs. Additionally, it can lead to uncomfortable temperature variations throughout your home. Evaluate existing insulation by assessing the current insulation in your home. Look for gaps, worn-out insulation, or areas with no insulation. Based on your assessment, select the appropriate type of insulation, with common options including fiberglass, cellulose, and spray foam insulation, with your choice depending on your budget and specific needs.

Insulation should be installed correctly to maximize effectiveness; hiring a professional is advisable if you're unsure about the installation process. In addition to insulation, sealing air leaks is crucial for maintaining a well-insulated home, with common areas for air leaks including doors, windows, and electrical outlets. Don't forget to insulate your attic as it's a significant source of heat gain in the summer and heat loss in the winter. By upgrading your insulation, you're making your home more energy efficient and reducing the workload on your HVAC system, which can translate to substantial savings in the long run.

Installing Carbon Monoxide Detectors: Safety Assurance

Safety should always be a top priority for your HVAC system. Carbon monoxide (CO) is a colorless, odorless gas produced by malfunctioning gas furnaces, water heaters, and other appliances in your home. Inhaling even small amounts of CO can lead to serious health issues or, in extreme cases, be fatal. To ensure the safety of your household, installing carbon monoxide detectors is a proactive step you should take. Install CO detectors on each level of your home, near sleeping areas, and close to your HVAC system and gas-burning appliances. Choose between battery-powered and hardwired CO detectors, with hardwired detectors typically connected to your home's electrical system, ensuring constant monitoring. Test your CO detectors monthly to ensure they function correctly by following the manufacturer's instructions. If your detectors are battery-powered, replace the batteries annually or when the low-battery warning sounds. If a CO detector sounds, evacuate your home immediately and call emergency services without re-entering until it's safe to do so. Installing carbon monoxide detectors is a simple yet essential proactive measure to protect your family from the potential dangers associated with your HVAC system.

Educating Family Members: Safety and Awareness

Your efforts to prevent future problems with your HVAC system shouldn't stop at maintenance and installation. Educating your family members about safety and awareness is equally important as your HVAC system is a significant part of your home, and everyone should be aware of its operation, potential hazards, and safety protocols. Regularly conduct safety briefings to inform your family about potential hazards related to the HVAC system, including carbon monoxide leaks, hot surfaces, and moving parts. Ensure that everyone in your household knows what to do in an HVAC-related emergency, such as a gas leak or CO detection. Teach family members how to operate the thermostat, adjust settings, and recognize common error codes on the HVAC system. Encourage your family to report unusual smells, noises, or temperature variations as timely reporting can help in addressing potential issues. Ensure your family knows the preventive maintenance schedule and the importance of inspections and filter replacements. Educating your family members creates a safety-conscious environment, ensuring everyone is prepared for any HVAC-related situation.

Proper maintenance is the cornerstone of HVAC longevity. In this chapter, we will delve into the nitty-gritty of systematic maintenance, educating homeowners about the nuances of HVAC usage, monitoring efficiency through performance metrics, considering energy-efficient components, adopting environmentally responsible practices through recycling and disposal, and knowing when it's time for a system replacement.

Systematic Maintenance: The Key to Longevity

Imagine your HVAC system as a finely tuned orchestra, with various components working harmoniously. To ensure this symphony continues to play smoothly, regular maintenance is crucial. Neglecting maintenance can lead to inefficiency, breakdowns, and, eventually, a shorter lifespan for your system. HVAC systems are not 'set it and forget it' appliances. Regular maintenance is the foundation for their longevity. Here are some essential aspects of systematic maintenance: Regularly scheduled maintenance check-ups are a must. Most HVAC professionals recommend having your system serviced at least once a year.

These check-ups involve thoroughly inspecting the system, cleaning, and necessary repairs. Your HVAC system's air filters maintain air quality and system efficiency. Clogged or dirty filters can force your system to work harder, leading to premature wear and tear. Depending on the type of filter, you might need to replace it every one to three months. Clean coils and ducts are essential for optimal heat exchange and airflow. Accumulated dirt and debris can hinder efficiency and potentially damage the system. Regular cleaning can prevent this. Moving parts in your HVAC system require proper lubrication.

Lubricating these parts can reduce friction, minimize wear, and help the system run smoothly. Faulty electrical components can pose a significant threat to your HVAC system. A regular inspection of electrical connections and components can prevent costly breakdowns. Fine-tuning your HVAC system can improve its performance and energy efficiency. This includes adjusting the thermostat, ensuring proper air balance, and verifying refrigerant levels. Your safety should always be a priority. A comprehensive maintenance check should include carbon monoxide testing, ensuring your system doesn't threaten your health.

Educating Homeowners: Proper HVAC Usage

Education is power; in the context of HVAC systems, it can make informed decisions and use your system effectively. Proper HVAC usage ensures a comfortable living environment and contributes to its longevity. Your thermostat is the command center of your HVAC system. Understanding how it works and using it efficiently can save both energy and money. For example, programming your thermostat to lower the temperature when you're not at home can lead to substantial energy savings. Proper air circulation is essential for an efficient HVAC system.

Keep vents and registers free from obstructions, ensuring air flows freely. Also, consider using ceiling fans to help distribute air evenly throughout your space. If your home has zoning systems, learn how to use them effectively. Zoning allows you to customize the temperature in different areas of your home, which can lead to significant energy savings. Educate yourself and your family about simple preventative measures, such as closing curtains or blinds during the hottest parts of the day to reduce the load on your air conditioner.

While routine professional maintenance is crucial, you can also monitor your system's performance. If you notice any irregularities, such as odd sounds, reduced airflow, or temperature fluctuations, don't ignore them as these can be early warning signs of issues that, if addressed promptly, can prevent major breakdowns. Using your HVAC system optimally can lower energy consumption, reduce wear and tear, and increase its lifespan.

Tracking Performance Metrics: Monitoring Efficiency

In today's data-driven world, monitoring the efficiency of your HVAC system is easier than ever. Knowing your system's performance can help you identify potential issues and optimize its operation.

Here are some important performance metrics to consider: Your energy bill is one of the most tangible indicators of your system's efficiency. A sudden spike in energy costs might signify an issue with your HVAC system, such as a clogged filter or refrigerant leak. The Energy Efficiency Ratio (EER) measures how efficiently your air conditioner operates when the outdoor temperature is at a specific level, with higher EER values indicating better efficiency. You can find this information in your system's documentation.

Seasonal Energy Efficiency Ratio (SEER) is another efficiency metric considering the varying outdoor temperatures throughout the year, with modern HVAC systems often having a SEER rating where a higher SEER number indicates greater efficiency. Annual Fuel Utilization Efficiency (AFUE) is relevant for heating systems as it measures the efficiency of a furnace or boiler, with a higher AFUE percentage meaning more efficient operation. An efficient HVAC system should maintain a consistent temperature throughout your home so monitor temperature variations between rooms - if some rooms are significantly warmer or cooler than others, your system might not be operating optimally.

Considering Upgrades: Energy-Efficient Components

As technology advances, so do the components and systems available for HVAC. Upgrading to energy-efficient components can enhance the performance of your HVAC system while reducing energy consumption.

Here are some upgrades to consider: If your heating system is aging, consider upgrading to a high-efficiency furnace or heat pump as these systems are designed to maximize heating performance while using less energy. Investing in a programmable thermostat can help you fine-tune your HVAC system's operation by setting specific temperatures for different times of the day and night, leading to substantial energy savings.

If it's time to replace your air conditioner, opt for an energy-efficient model with a high SEER rating as this investment can pay off through lower energy bills and a longer system lifespan. Improving your home's insulation and sealing gaps and cracks can reduce the load on your HVAC system so that when your home is well-insulated, your HVAC system doesn't have to work as hard to maintain a comfortable temperature. Many modern HVAC systems can be integrated with smart home technology, allowing you to control your HVAC remotely and optimize your system's operation more easily. When considering upgrades, consulting with a professional HVAC technician is essential as they can guide you on compatible components with your existing system that offer the best return on investment.

Recycling and Disposal: Environmentally Responsible Practices

Responsible disposal of HVAC components is not only good for the environment but also for your community since many HVAC systems contain materials that can be harmful if not disposed of properly.

Here's how to approach recycling and disposal responsibly: Older HVAC systems may contain refrigerants that are harmful to the environment so when replacing such systems, it's essential to have a certified technician recover and recycle the refrigerant. Many components of HVAC systems, including metal parts and wiring, can be recycled so check with local recycling facilities when replacing or disposing of your HVAC system to ensure recyclable materials are appropriately handled. For components or systems that can't be recycled, such as certain types of insulation or air filters, follow local regulations for their disposal as many areas have specific guidelines for disposing of hazardous materials. In some cases, HVAC components that are still in good condition can be donated or repurposed so organizations or individuals needing HVAC parts can benefit from your old system's components. When hiring HVAC professionals for installation or replacement, ensure they have environmentally responsible practices including proper refrigerant handling and disposal procedures. By adopting environmentally responsible practices in the recycling and disposal of HVAC components, you contribute to a cleaner planet and potentially benefit from local recycling programs and goodwill in your community.

Planning for System Replacement: When It's Time

No matter how well you maintain your HVAC system, there comes a time when replacement is the only viable option. While it might be challenging to say goodbye to a trusted system, replacing it at the right time can save you money and increase efficiency.

Here are some signs that it's time for a system replacement: HVAC systems typically have a lifespan of 10 to 15 years for air conditioners and 15 to 20 years for furnaces so if your system surpasses these age limits, replacement is worth considering. If you call for HVAC repairs more often than not, investing in a new system might be more cost-effective since frequent repairs can quickly add up. As your system ages, it becomes less efficient so you may notice higher energy bills and reduced performance and replacing the system with a more efficient one can save you money in the long run.

If your HVAC system struggles to maintain a consistent temperature throughout your home, it's a sign that it might be on its last legs since newer systems can provide better climate control. Advancements in HVAC technology can significantly improve system efficiency so upgrading to a newer model can provide better features and energy savings. If your household has grown or if you've made significant changes to your home, your existing HVAC system might not meet your needs, and in such cases, a replacement or system upgrade is often necessary.

PART 3

COMFORT AND ADVANCED TECHNIQUES

PART 3

CHAPTER 5

Indoor Air Quality and Comfort

Achieving optimal indoor air quality and comfort should be a top priority for any home or building owner. The air we breathe indoors greatly impacts our health, productivity, and well-being. Poor indoor air quality from pollutants or improper humidity levels can cause respiratory issues, allergies, headaches, and more. Furthermore, an uncomfortable indoor environment leads to dissatisfaction and lower performance for occupants. This chapter will explore the technologies and best practices for enhancing indoor air quality and comfort through proper HVAC system design, maintenance, and operation. First, we will discuss why indoor air quality matters and review different air filtration methods. Next, we will cover recommendations for ideal humidity levels and humidity control strategies. We will also examine HVAC zoning techniques for improving temperature and humidity control in different areas of a home or building. Finally, we will guide balancing comfort and energy efficiency to create a pleasant, healthy, efficient indoor environment. Follow these recommendations to maximize air quality and comfort for those spending time in your home or facility.

PRIORITIZING INDOOR AIR QUALITY: WHY IT MATTERS

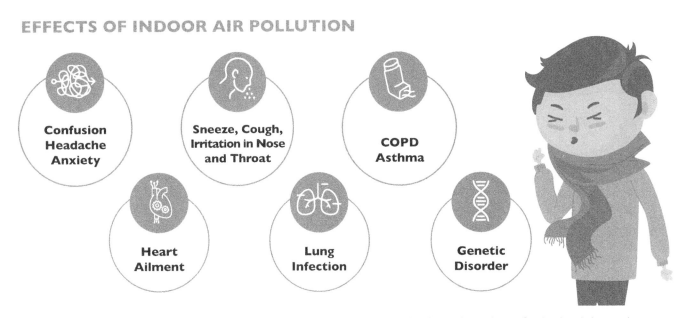

Indoor air quality refers to the cleanliness and pollution levels of air inside a home, office, school, or other interior space. While outdoor air quality receives significant attention, the air inside buildings and structures can be far more contaminated and hazardous due to tight seals, lack of air circulation, and indoor pollutants. Poor indoor air quality can occur in any building but is especially prevalent in tightly sealed, energy-efficient structures. All homeowners and facility managers should make indoor air quality a top priority. Failing to address indoor air pollution can lead to acute and chronic health issues ranging from irritation to respiratory disease. Improving air quality generates countless benefits for long and short-term health, productivity, and well-being.

Health Effects of Poor Indoor Air Quality

Poor indoor air quality introduces a wide array of health risks. Most immediately, high pollutant levels can cause eye, nose, and throat irritation, making occupants uncomfortable. Headaches, fatigue, shortness of breath, sinus congestion, and coughing are common side effects of contaminated indoor air. More severely, long-term exposure can increase the risks of asthma, allergies, cardiovascular disease, and certain cancers. Volatile organic compounds or VOCs off-gassed from building materials, furnishings, and cleaning chemicals have been linked to liver, kidney, and central nervous system damage. Biological pollutants like mold, dust mites, pet dander, and cockroach droppings can trigger allergic reactions and asthma episodes in sensitive individuals. Carbon dioxide build-up leads to lower cognitive function and difficulty concentrating without proper ventilation and filtration. Children, seniors, and those with existing conditions are most vulnerable to indoor air pollution's health effects. However, even healthy adults face increased chances of developing chronic diseases from prolonged exposure.

Productivity and Performance Impacts

Studies show poor indoor air quality substantially hampers productivity, performance, and satisfaction. Office workers in environments with high VOC levels commit nearly 60% more errors in cognitive function tests. Students in classrooms with elevated CO2 and pollutants receive lower scores on standardized tests. Contaminants like formaldehyde, benzene, nitrogen dioxide, and ozone are proven to negatively impact decision-making, concentration, and response time. Symptoms from allergies, headaches, fatigue, and irritation also make working or studying difficult. Enhanced ventilation and air filtration have increased productivity among office workers by as much as 11%. Meanwhile, high indoor pollutants decrease retail shopper's willingness to purchase by up to 90%. For commercial facilities, prioritizing air quality leads to happier, healthier, and higher-performing occupants, along with increased sales and profits.

SOURCES OF INDOOR AIR POLLUTION

- Chemicals from Building Materials
- Animal dander and Hair
- Mold & Bacteria
- Chemicals from Paint and VOCS
- Gases from Fireplace
- Outdoor Air Pollutants
- New Electronics and Broken Lighs
- Cigarette Smoke
- Cleaning Supply Chemicals
- Carbon monoxide
- Chemicals seeping through foundation

Sources of Indoor Air Pollution

To improve air quality, it is essential to identify sources of indoor pollutants. Outdoor air enters buildings with common pollutants like particulate matter, ground-level ozone, vehicle exhaust, and pollen. However, studies show indoor levels of many contaminants exceed outdoor levels even in highly polluted cities. This is because the tightly sealed nature of modern buildings traps pollutants while many indoor sources generate new hazards. Off-gassing from construction materials, furniture, carpeting, and cleaning supplies releases VOCs, including formaldehyde and benzene.

Mold growth fueled by moisture or water damage generates spores and mycotoxins. Pests like cockroaches and rodents spread allergens and bacteria. Everyday activities like cooking, cleaning, and even breathing introduce particulate matter, VOCs, carbon dioxide, and humidity into the air. Without constant ventilation and filtration, pollutants accumulate and mix, forming an increasingly toxic indoor air quality environment.

Populations Vulnerable to Indoor Air Pollution

While poor indoor air quality affects everyone, certain populations face higher risks for health effects. Infants, children, and older people tend to be most vulnerable due to developing or compromised respiratory and immune systems. Those with existing conditions like asthma, cardiovascular disease, or allergies also suffer exacerbated symptoms when exposed to common indoor irritants. Additionally, people who spend much time indoors, like office workers, homemakers, infants, and seniors, have more opportunities for prolonged contact with contaminants. However, indoor pollution also impacts healthy adults through increased risks for disease later in life. While controlling personal risk factors is advisable, providing optimal indoor air quality through building management benefits all occupants.

Benefits of Enhanced Indoor Air Quality

Improving indoor air quality generates immediate and long-term benefits for comfort, health, cognitive function, and productivity. By removing pollutant sources and increasing ventilation, irritation, allergy, and asthma symptoms substantially decrease. Respiratory infections like colds, flu, and bronchitis also decline thanks to reduced contact with infectious aerosols. With lower pollutants, cognitive abilities improve along with concentration, decision-making, and reaction times. Studies document enhanced productivity and work satisfaction after indoor air quality interventions in offices. At home, better air quality means better sleep, improved energy, and more quality time for family. Increasing indoor air quality helps reduce the burden of respiratory disease and chronic conditions for society. While upfront investments in infiltration and ventilation may be required, enhanced indoor air quality benefits outweigh the costs over time.

AIR FILTRATION AND PURIFICATION TECHNOLOGIES

The most effective way to improve indoor air quality is source control and ventilation. However, air filtration and purification technologies provide supplemental removal of hazardous particulate and gaseous pollutants. From simple filters to advanced purifiers, today's options target various contaminants with varying effectiveness and cost.

Filter Fundamentals

Air filters provide a first defense against particulate indoor air pollution like dust, pollen, and mold. By trapping particles as air passes through the HVAC system, filters stop contaminants from circulating and accumulating indoors. They also protect HVAC components from dust buildup and damage. The most common are low-efficiency fiberglass or foam panel filters with a MERV rating of 1-4. While inexpensive, these only remove large particles over 10 microns in size. Look for extended surface pleated filters with MERV ratings above 8 for superior filtration. These capture much smaller particles down to around 3 microns. Finally, HEPA filters achieve MERV 17-20 ratings by removing over 99% of particles above 0.3 microns, including dust, pet dander, mold, and smoke. Upgrade your HVAC air filters to capture more particulate pollutants from supply air.

Ionizing and Ozone Air Purifiers

Some air cleaning devices use ionization or ozone generation to neutralize indoor pollutants. However, neither method removes the hazard from the air. Ionizers electronically charge air molecules and particulates to make them cling together and settle out faster. But charged particles remain to be inhaled. Ozone generators help eliminate odors, mold, and VOCs by producing ozone gas. However, ozone does not fully clean air and, at higher concentrations, causes significant respiratory irritation and long-term lung damage. Despite marketing claims, ionizing and ozone air purifiers provide minimal benefits for indoor air quality. Avoid these devices favoring source control, ventilation, and proven filtration technologies.

Activated Carbon Filtration

While filters trap particulates, air purifiers with activated carbon adsorb gaseous pollutants. As air passes over the carbon, VOCs, odors, and some biological contaminants adhere to the highly porous material. Look for systems with large carbon beds and frequent filter replacements to maintain absorption capacity. Portable purifiers with activated carbon substantially reduce VOCs and odors in single rooms. At the whole building level, air handlers with carbon filters lower overall concentrations of radon, auto exhaust, and off-gassing chemicals. Activated carbon effectively complements particulate filtration to improve the removal of both solid and gaseous indoor air pollutants.

Ultraviolet Germicidal Irradiation

Ultraviolet germicidal irradiation or UVGI air purifiers inactivate biological contaminants with short wavelength UV-C light. UV exposure by damaging DNA and RNA kills or deactivates microbes like viruses, bacteria, and mold traveling through HVAC systems. UVGI is highly effective against pathogens but does not remove other particulate or gaseous pollutants. Systems installed in ductwork provide whole-building disinfection, while portable UV units work well in single rooms. Use high-output UV lamps designed for germicidal use and follow all safety precautions. While not a standalone solution, UVGI is a powerful supplement to filtration for reducing bioaerosols, including infectious particles.

Photocatalytic Oxidation Purifiers

Photocatalytic oxidation or PCO air purifiers combine UV light, a titanium dioxide catalyst, and activated carbon to eliminate VOCs rather than absorb them fully. As contaminated air passes through the unit, UV light energizes the catalyst to trigger an oxidation reaction that breaks down VOCs into carbon dioxide and water vapor. The carbon bed mops up any remaining VOCs. While costly, PCO purifiers destroy indoor air VOCs without generating harmful byproducts. However, they do not remove particulate matter. PCO systems are highly effective against volatile indoor air pollutants like off-gassing chemicals.

Whole-Building vs Portable Purifiers

Air purifiers are available as in-duct systems to clean a building's full air supply or as portable units for single-room use. Whole-building options provide comprehensive air quality but require professional installation and maintenance. Portable HEPA and activated carbon purifiers effectively clean individual spaces for rental properties or partial renovation. Look for high CADR ratings for particulate and chemical reduction. Size portable units appropriately for room footage and follow all placement recommendations. Combine centralized and portable purifiers to bring both whole-building and local indoor pollutants under control economically.

ACHIEVING IDEAL HUMIDITY LEVELS FOR COMFORT

Indoor humidity impacts air quality, occupant comfort, and HVAC system efficiency. Keeping indoor relative humidity or RH within an optimal 40-60% range minimizes mold growth, dust mites, and other biological contaminants while creating comfortable environmental conditions. Humidity levels that are too high or too low lead to issues ranging from stuffy, stale air to dry eyes, cracked skin, and increased static electricity.

Impacts of Improper Humidity Levels

Excessively high or low indoor relative humidity causes numerous problems for air quality, comfort, and health. Humidity over 60% encourages the proliferation of mold, dust mites, bacteria, and viruses, leading to allergies, asthma, and respiratory issues. Dampness also contributes to mechanical issues like door or window swelling. Low humidity under 40% results in irritation like dry eyes, cracked skin, sinus congestion, and increased susceptibility to colds and flu. Extremely dry air also causes static electricity issues and damage to wood furnishings. For optimal comfort and well-being, indoor relative humidity should be maintained between 40-60% as much as possible.

Recommended Humidity Setpoints by Season

Ideal indoor humidity setpoints vary somewhat by season to account for changes in outdoor climate and human physiological needs. In winter months, aim for 30-40% relative humidity. This helps counteract exceedingly dry, heated air while minimizing window condensation issues. Setpoints are slightly to 40-50% RH during spring and fall to match the moderate transitional outdoor conditions. A humidity of 50-60%

is recommended for summer to feel pleasantly cool without being damp. Dehumidify excessively humid basement, crawl space, and closet areas to 60% RH or less year-round. Adjust thermostat humidity settings seasonally and use portable dehumidifiers as needed in damp zones.

Devices for Measuring Humidity

Using reliable measurement devices, you cannot effectively control indoor humidity without first monitoring levels. Simple humidity meters provide constant spot checks and identify problem areas. Look for accuracy within 2-5%. Data logging hygrometers give ongoing RH readings to determine if levels are too high or low over time. More advanced options like psychrometers directly measure dewpoint while building automation systems have humidity sensors giving whole building data. Use several strategically placed measurement devices to get complete moisture profiles. Replace sensors every 2-3 years for continued accuracy. Without proper humidity measurement, you cannot verify if the relative humidity is within the ideal comfort range.

Managing Humidity Through HVAC Equipment

Most humidity control occurs through proper HVAC system sizing, operation, and maintenance. Heating and air conditioning removes moisture from the air as a natural byproduct of temperature change. Systems must be sized correctly to handle latent loads. Increase runtimes to remove additional moisture if needed. Sealing ductwork prevents humid interior air from entering and overtaxing the system. Clean evaporator coils allow proper dehumidification. Monitor that condenser coils discharge all captured moisture outside. Keep air filters clean for airflow efficiency. Ensure adequate ventilation brings inappropriate levels of fresh, dry air. Humidity problems frequently stem from underlying HVAC issues.

Supplemental Humidity Control Strategies

Install dedicated mechanical equipment like whole-home dehumidifiers or humidifiers for supplementary humidity control. Use portable units in especially damp rooms like basements and bathrooms. Dehumidifiers work more effectively than air conditioning for moisture removal. Active humidifiers properly maintain RH in dry climates, while passive options like water pans suffice for short-term relief. Ventilate kitchens and bathrooms during and after use to remove moisture at the source. Avoid oversized humidifiers, which raise humidity too high. Remove large moisture sources like indoor pools or limit evaporative coolers in humid climates. Use moisture sorption products to absorb excess humidity passively. Manage indoor humidity through both HVAC improvements and supplemental moisture control.

ZONING YOUR HVAC SYSTEM: ROOM-BY-ROOM CONTROL

Implementing an HVAC zoning system is one of the most effective ways to improve temperature and humidity control. Zoning divides your home or building into distinct areas with separate environmental regulations based on usage and occupancy patterns. While zoning requires an increased upfront investment, the long-term benefits of comfort and efficiency make it worth the cost for most structures.

Benefits of Zoned HVAC Control

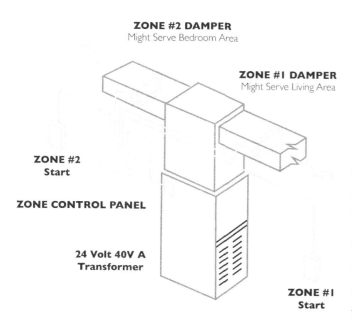

Zoning delivers more tailored heating and cooling to match different rooms' utilization. Rarely occupied rooms can be kept at less extreme setpoints to save energy. Meanwhile, comfort improves in gathering areas like living rooms where conditions are optimized. Zones reduce issues with certain spaces being too hot or cold due to their location and orientation. Individual moisture control reduces humidity variability to minimize mold and mildew growth. Zone controllers enable customized schedules for nights, weekends, and vacation periods. Advanced zoning systems even allow voice-activated commands or management from smartphones and other devices. Zoning also increases HVAC system longevity by eliminating unnecessary runtime in unused rooms. For both improved comfort and efficiency, zone control is a smart investment.

Designing a Zoning Layout

When designing an HVAC zoning strategy, divide areas with significant occupancy, temperature, or humidity differences into separate zones. Frequently used spaces like living rooms and kitchens get dedicated zones, while rarely occupied rooms can be grouped. Areas exposed to intense solar gain, like west-facing master bedrooms, get their zones to counteract afternoon overheating. Separate basements and top-floor rooms from main living levels to address their temperature extremes. Give each bedroom its zone if possible so occupants can customize conditions. For humidity, isolate notoriously damp zones like crawlspaces, bathrooms, and laundry rooms for moisture control. Create as many zones as your budget allows for maximum comfort control.

Dampers for Low-Cost Zoning

The simplest way to achieve basic zoning is manually adjusting dampers on branch ducts feeding different areas. Open or close dampers to direct more or less airflow to each zone. While requiring manual adjustment, dampers provide zone separation for under $100 in materials. Motorized dampers can automatically open or close based on a schedule or thermostat input. Retrofit existing ducts with dampers or install dampers during new ductwork installation. Use dampers to create separate zones for major problem areas like a remote main suite. However, dampers alone provide limited environmental control and lack automation benefits.

Zone Control Boards and Dampers

For advanced automated zoning, install a zone control board with motorized dampers. The control board contains damper actuators and thermostats, allowing each zone to regulate temperature independently. Some systems integrate humidity control and ventilation needs as well. Occupants can customize schedules and setpoints room-by-room. Advanced zone controllers cost around $300-$800, with several hundred dollars more per zone for damper installation. Zone control boards also work with newer HVAC systems with variable speed compressors and blowers for fine comfort tuning. While an investment, they deliver superior comfort and efficiency.

Multi-Stage and Variable Equipment

An alternate advanced zoning strategy is installing multi-stage or variable-capacity HVAC equipment capable of modulating output to match lighter or heavier loads. For example, a two-stage unit runs at half capacity during mild weather and then ramps up to full output when needed. Variable speed systems offer even more gradation for smooth runtime and temperature control. While not true zoning, this achieves some efficiency and comfort benefits without extensive damper.

Mini-Split Zoning Systems

One popular emerging option is mini-split heat pumps, which achieve zoning capabilities through multiple indoor evaporator units connected to an exterior central condenser. Ductless, high-efficiency mini-splits allow different rooms to be heated or cooled independently based on individual thermostat control. Standard configurations utilize one outdoor unit with 2-4 indoor heads. Additional heads can be added for whole building coverage across multiple zones. Installation costs more than traditional splits, but lower ductwork needs to offset this premium. Consider mini-splits when ductwork is inaccessible or existing HVAC lacks capacity for zoning.

Designing and operating HVAC systems inherently requires tradeoffs between comfort and energy efficiency. Most comfort desires, like colder temperatures, uniform conditions, and constant air circulation, require additional energy. However, comfort and efficiency do not have to be mutually exclusive goals. With careful system design, control strategies, and active management, you can strike an optimal balance tailored to your needs and priorities.

Design Factors for Balancing Comfort and Efficiency

Several design decisions significantly impact the balance between comfort and efficiency. Oversizing equipment reduces humidity control, causes short cycling, and increases costs. Proper sizing with room for expansion improves moisture removal and savings. Specifying high-efficiency components like ECM motors cuts energy use. Include zoning capabilities to optimize conditions in occupied spaces. Orient ducts towards sunny areas require more cooling. Install larger ducts with less restrictive airflow to rooms with greater loads. Include programmable thermostats and advanced controls for smart optimization. Model energy savings from efficiency upgrades to justify costs. Design choices establish the system's comfort and efficiency parameters.

Advanced Equipment for Enhanced Control

New HVAC technologies allow greater precision, balancing comfort and efficiency. Two-stage or variable capacity systems modulate output to match loads precisely. Variable speed components like ECM blower motors adjust airflow and reduce energy use. Zone damper systems direct conditioning to where it is needed most. Smart thermostats finetune setpoints hour-by-hour based on occupancy. Demand response capabilities allow cooperation with utility rebate programs. While premium equipment carries a cost, expanded controllability optimizes indoor conditions while conserving energy.

Control Strategies to Harmonize Comfort and Efficiency

Advanced control programming allows HVAC systems to achieve both comfort and efficiency simultaneously. Setback temperatures lower heating and raise cooling setpoints during unoccupied periods to conserve energy. Optimum start algorithms initialize equipment before occupancy so spaces are comfortable upon arrival. Adaptive temperature reset slightly adjusts thermostat setpoints up or down based on how quickly zones heat up or cool down. Humidity sensors maintain ideal moisture levels without overcooling. Adjust control sequences to match actual occupancy patterns for maximum comfort and savings.

Operational Techniques for Balancing Comfort and Efficiency

How HVAC systems are operated daily strongly impacts comfort, costs, and longevity. Perform regular maintenance, like replacing filters, to sustain performance. Reset systems for unoccupied periods using programmable schedules. Pre-cool spaces before peak afternoon hours to reduce load. Turn off or isolate unused spaces. Provide access to thermostats and educate occupants about energy-saving behaviors. Relaxing dress codes expands the comfort zone and reduces HVAC runtime. Avoid extreme setups to limit simultaneous heating and cooling. Operational discipline improves comfort satisfaction while controlling expenditures long-term.

Occupant Engagement for Achieving Goals

Engaging with occupants is key to balancing comfort and efficiency in shared buildings. Gather feedback to understand problem areas and priorities. Communicate energy reduction goals and the role occupants play in supporting those efforts. Provide options for individual comfort control like fans, desk heaters, and flexible clothing policies. Design easily accessible and intuitive zone thermostats to encourage participation. Recognize occupants who embrace conservation behaviors. Address major comfort complaints immediately. Include occupants as partners in optimizing their environment while helping enhance building performance.

Commissioning for Quality Assurance

Commissioning is a quality assurance process that optimizes new HVAC systems for comfort and efficiency. Commissioning verifies components are installed properly, functional as designed and correctly integrated. Operational sequences are tuned to match specifications. Staff are trained in operating the system efficiently. Deferred functional testing after initial occupancy ensures comfort and savings goals are sustainably achieved. While adding 5-10% to costs, commissioning avoids comfort issues, energy waste, and premature repairs. For new construction and retrofits, commissioning brings occupant satisfaction and performance in alignment.

Chapter 6

Advanced HVAC Techniques and Technologies

Heating, ventilation, and air conditioning (HVAC) systems have come a long way from the early days of basic heating furnaces and window air conditioning units. Today's HVAC systems incorporate advanced technologies that provide superior efficiency, sustainability, controls, and performance. By leveraging innovations like geothermal heating and cooling, solar power, variable refrigerant flow, and energy recovery ventilation, HVAC systems can be finely tuned to provide the highest comfort level at the lowest operating cost.

This chapter will explore some of the most cutting-edge and promising HVAC technologies available today and on the horizon. We will examine how these systems work, their advantages over traditional HVAC, real-world applications and case studies, installation considerations, maintenance best practices, and cost analyses. Our goal is to provide readers with an in-depth look at the next generation of HVAC so they can make informed decisions about upgrading their systems to maximize efficiency and savings over the long term. The topics covered include:

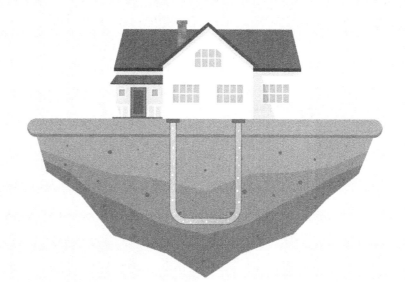

Geothermal HVAC (heating, ventilation, and air conditioning) systems, also known as ground source heat pump systems, utilize the earth's natural temperature for heating and cooling. These systems take advantage of the fact that the earth's temperature remains relatively constant year-round below the surface. A geothermal system uses a ground heat exchanger to transfer heat between the relatively stable underground temperatures and the building being heated or cooled. In the winter, heat from the ground is absorbed and concentrated to warm the building. In the summer, heat is extracted from the building and dispersed into the cooler earth.

Geothermal systems have major advantages over conventional HVAC systems that rely on outdoor air temperatures. They can achieve 300-600% efficiencies compared to 80-140% for air source heat pumps because they draw from the earth's vast reservoir of underground solar energy. The temperatures underground are more moderate and stable than above-ground air temperatures. This allows geothermal systems to provide consistent heating and cooling year-round, unaffected by extreme outdoor temperatures. The systems are also not dependent on fossil fuels and have minimal greenhouse gas emissions.

Types of Geothermal Heat Pumps

Two main types of geothermal heat pump systems are used for HVAC: closed loop and open loop. Closed loop systems are the most common and recirculate an antifreeze solution through underground pipes. Open-loop systems pump groundwater directly from a well for heat transfer. Three main subsurface designs within closed-loop systems exist vertical boreholes, pond loops, and horizontal trenches. The optimal design depends on the climate, soil type, land availability, and local installation costs. All geothermal systems use a heat pump to concentrate heat from the ground and disperse it into the building's air distribution system.

Vertical closed-loop systems

Vertical closed-loop systems are the most commonly used geothermal design in residential applications. They involve drilling boreholes 150 to 300 feet deep into the ground using a drilling rig. Then, a pair of polyethylene U-shaped loops are inserted into each borehole. The loops are connected with a horizontal pipe to complete the circuit. A 30% propylene glycol antifreeze solution is pumped through the pipes to transfer heat to and from the ground. The small diameter of the vertical loops allows for effective heat transfer, given the greater temperatures found deeper underground. This makes vertical loops well-suited for heating-dominated climates. The upfront cost of vertical drilling can be high but requires less land space than horizontal loops.

Horizontal closed-loop geothermal systems

Horizontal closed-loop geothermal systems are more cost-effective for cooling-focused climates that only require shallower underground piping. These systems use a backhoe or trencher to dig 3 to 6 feet deep long trenches. Polyethylene piping loops are laid in the trenches and connected to form a continuous circuit. The trenches are backfilled once the pipes are placed. More yard space is needed than vertical loops, but trenching is cheaper. Larger commercial buildings often use horizontal loop fields with hundreds of trenches. If available, the loops can also be submerged in an existing pond or lake.

Pond closed loop systems

Pond closed loop systems, also called surface water heat pumps, are an alternative to buried underground loops. Long sections of coiled polyethylene tubing are submerged at least 8 feet below the surface of a pond, lake, or reservoir on the property. The coils should cover 1/4 to 1/2 acre of surface area. Care must be taken to prevent buildup of biological organisms that could clog the coils. A sediment trap and chemical treatment may be necessary. A water well can also function as a vertical pond loop. Surface water sources offer good heat transfer and are easier to maintain than buried loops. But they depend on an existing body of water.

Open-loop geothermal systems

Open-loop geothermal systems pump and discharge water from onsite water wells for direct heat transfer versus a closed underground loop. The system draws water from one well, circulates it through a heat exchanger, and discharges it into a second infiltration well, pond, or other approved location. This provides an abundant heat source but depends on accessible groundwater. Open loops are common where there are sufficient groundwater and injection discharge options. They eliminate the antifreeze solution and buried piping of closed loops. However, the water quality must be monitored, and disposal options must be maintained over time.

Geothermal Heat Pump Operation

Geothermal heat pumps operate on the simple principle of using the earth's natural heat for heating and cooling. All geothermal systems have three main components: the ground heat exchanger, heat pump unit, and air delivery system. The ground heat exchanger consists of pipes called loops that are buried underground. A water and antifreeze solution circulates through these pipes to absorb or reject heat from the ground. The heat pump unit includes a compressor, evaporator, expansion valve, and condenser coil. It manipulates refrigerant pressures and the phase change between liquid and gas to concentrate ground heat in winter or extract building heat in summer. The air delivery system uses ductwork and vents to circulate conditioned air throughout the building.

In heating mode, the liquid refrigerant absorbs heat from the cooler ground loops passing through the evaporator. It evaporates into a gas, then gets compressed to a higher temperature gas that transfers heat into the building's air stream via the condenser. This heat originates from the stored solar energy in the ground. The refrigerant is expanded and condensed into a liquid to repeat the cycle. So, the heat pump takes the roughly 50F to 60F temperatures of the earth and raises them to 95F+ for heating.

In cooling mode, the process is reversed - the refrigerant absorbs heat from the warmer indoor air stream via the evaporator. It releases it into the cooler ground loops through the condenser. This transfers building heat into the earth, where it can dissipate given the enormous heat sink capacity of the ground. The heat pump enables heating and cooling from a free, renewable geothermal resource.

Examining the refrigeration cycle in more detail - the working fluid starts as a saturated liquid refrigerant going into the evaporator, which boils and vaporizes after absorbing heat from the ground loops. The low-pressure vapor refrigerant then enters the compressor, increasing pressure and temperature. The high-pressure superheated vapor enters the condenser and condenses back into a liquid state while releasing heat to the building's air distribution ducts.

The high-pressure liquid passes through an expansion valve, which drops the pressure and temperature before re-entering the evaporator to start the cycle again. This closed-loop refrigeration process efficiently moves heat from the moderate ground to warm the building in winter or vice versa in summer. Modern variable-capacity heat pumps optimize the operating pressures and temperatures year-round.

Unlike closed-loop systems, open-loop geothermal heat pumps work by directly pumping groundwater from a supply well through a heat exchanger inside the heat pump unit itself. The groundwater absorbs or deposits heat via the refrigerant-to-water heat exchanger, depending on the heating or cooling mode. The water is then disposed of in an injection well or other approved method. No subsurface piping loops are needed. The constant groundwater flow provides excellent heat transfer and availability while eliminating antifreeze solution maintenance.

Some open or pond loop geothermal systems use direct geo-exchange instead of a heat pump. The building's distribution ducts circulate refrigerant directly in a closed loop between the air handler and the ground source. This eliminates the intermediate heat exchanger and simplifies the system. However, it provides less temperature control capability than a heat pump unit. Direct geo-exchange works best in moderate climates with smaller heating/cooling loads.

Real-World Applications and Examples

While geothermal systems involve a higher upfront investment than conventional alternatives, the return on investment can be significant given their extreme efficiency and stable operating costs. Geothermal HVAC can be applied to all building types - single-family homes and large commercial structures. Some real-world applications and examples include:

For residential applications, geothermal heat pumps are an excellent option for homeowners looking to reduce energy costs while improving comfort. A typical single-family home uses vertical borehole closed loops for compact installation. A recent case study examined a 2,800-square-foot suburban Oklahoma home with an aging gas furnace and central air system. The owner replaced this with a 3-ton geothermal heat pump system with four 150-foot vertical boreholes.

The installed cost was $16,000, but the owner qualified for a 30% federal tax credit, bringing the net cost down to $11,200. Prior heating and cooling bills averaged $270 per month. With the new geothermal system, the monthly costs dropped to just $89, providing over $2,100 in annual savings. Factoring in the tax credit, the simple payback period for the new system was under 5 years. The homeowner reported much quieter and more even heating and cooling with the geothermal heat pump compared to the old system.

For commercial buildings, geothermal HVAC scales up effectively because of the modular nature of loop fields. A school or university campus provides an ideal large-scale geothermal case study. Ball State University in Indiana retrofitted its entire heating and cooling system to geothermal to reduce costs and emissions. The project included over 4,000 boreholes, two energy stations housing heat pumps and cooling towers, and a distribution piping network.

The $72 million project was partially funded by a $5 million federal grant and is now saving the university $2 million annually in energy costs. The converted system has cut the campus' carbon footprint in half. Geothermal provides even, reliable comfort year-round for students, faculty, and staff across dozens of buildings while showcasing the environmental and economic benefits. This project demonstrates geothermal's potential for large-scale deployment.

Commercial office buildings are another prime candidate for geothermal retrofits or new construction. A three-story, 51,000-square-foot office building in New York provides a great case study. The existing fossil fuel boilers and electric chillers were past their useful life. The owner opted to install vertical closed loop geothermal using 144 boreholes at 250 feet deep to accommodate the large heating and cooling loads.

The Category D office building previously had annual energy bills of $210,000. With the new geothermal system, this dropped to just $75,000, saving $135,000 in annual energy costs. The $1.2 million system cost was offset by $400,000 in grants and tax credits. The vastly improved efficiency, offset incentives and ongoing savings resulted in a sound 7-year ROI. Tenants also benefit from greater comfort. This demonstrates geothermal's financial case for large office buildings.

Large public facilities like libraries, hospitals, airports, and government buildings ramp up geothermal utilization based on its public image and operating benefits. For example, the Colorado Poudre Valley Public Library District wanted to expand its main location while reducing its environmental impact. They drilled 420 boreholes to construct an expansive closed-loop geothermal well field under the parking lot. The new HVAC system cut natural gas use by 70% and electricity by 30%.

With substantial utility rebates, the overall cost was comparable to conventional HVAC. The library now promotes its geothermal leadership and invites locals to tour the mechanical room. Geothermal supports its public education mission while slashing carbon emissions. This exemplifies how geothermal can benefit large public institutions and the broader community.

Mixed-use commercial developments can also harness geothermal at a large scale with multiple building types involved. For example, a multifamily housing complex in British Colombia with 238 residential units and 20,000 square feet of commercial space installed closed loop geothermal to heat and cool the entire property. One hundred fifty vertical boreholes provide heating, while pond loops sourced from an existing reservoir onsite provide summer cooling.

According to the developers, this hybrid geothermal system reduces utility costs by 60% compared to conventional HVAC. Substantial rebates and tax credits offset the upfront costs. Geothermal provides an ideal way to sustainably heat and cool a combination of residential apartments and commercial real estate. This highlights the technology's versatility and economies of scale.

Installation, Maintenance, and Cost Considerations

Proper installation, regular maintenance, and understanding of cost implications are key to maximizing a geothermal system's performance and return on investment. Climate, soil type, land availability, and local energy rates impact the feasibility and payback period. Critical steps include careful project analysis, design, equipment selection, and installation. Ongoing maintenance, like checking fluid levels and filtering air handlers, will prevent issues.

Performing heat load calculations for the building is an important first step in designing a geothermal system. The heating and cooling requirements and peak loads will dictate the ground loop heat exchanger size, pipe materials, fluid properties, and heat pump capacity needed. Proper ground loop piping material selection, layout, depth, and spacing also play a key role. HDPE and fusion welding are standard for most closed-loop pipe configurations.

A certified and experienced geothermal loop field installer is recommended - this is not typically a DIY project. The earth connection loop field is the most critical and permanent system component. Improper installation can lead to subpar heat transfer and higher long-term operating costs. Local codes must also be consulted regarding piping depth, borehole cementing, and other requirements.

The heat pump and air delivery equipment selection is the next key phase. Qualified HVAC professionals should size, specify, and integrate these components according to the calculated peak heating/cooling loads and loop heat exchanger design. Proper refrigerant pressure, flow rates, and temperature differential must be maintained for efficiency. The operating parameters are balanced through the heat pump controls.

Ongoing maintenance best practices for geothermal systems include annual checkups to inspect refrigerant levels, check loop fluid chemistry, and pressure tests for leaks. The air filters, coils, and vents should also be cleaned and serviced regularly. Some closed-loop systems may require fluid replacement every 5-10 years. Open loops and surface water loops require more monitoring of water quality and disposal systems.

Regarding upfront costs, geothermal systems cost more to purchase and install than conventional alternatives - typically around $6,000-$10,000 per ton of heating/cooling capacity. But this premium cost is recouped through dramatically lower operating expenses. With current tax credits and incentives, 3-7-year payback periods are common. Over a 25+ year lifespan, geothermal systems save tens of thousands in energy costs while increasing comfort and reducing carbon emissions.

SUSTAINABILITY AND SAVINGS: SOLAR-POWERED HVAC SYSTEMS

Solar-powered heating and cooling, also known as solar thermal HVAC, harnesses energy from the sun to provide heating, cooling, and hot water for buildings. Solar collectors absorb solar radiation to heat fluids that are then used to heat air or water. In cooling mode, solar heat drives double-effect absorption chillers. Solar PV systems can also power electric HVAC components like fans and pumps. Solar HVAC technologies offer a renewable energy alternative that reduces fossil fuel usage and environmental impact. The availability of solar rebates and tax credits in many regions also improves return on investment.

Types of Solar Thermal Collectors

Two main categories of solar collectors are used in HVAC systems: non-concentrating and concentrating. Non-concentrating collectors include flat plate and evacuated tube designs. Concentrating collectors use reflective surfaces to focus sunlight onto a small absorber area. Common types are parabolic troughs and solar dishes. The optimal solar collector depends on the climate and application. Flat plates are the simplest and most affordable, while concentrating designs offer higher efficiencies in certain weather conditions.

Flat plate solar collectors consist of a dark flat plate absorber covered in glass or plastic and insulated on the back. The absorber heats up when sunlight hits and transfers the heat into fluid-filled tubes attached to the plate. They are simple, durable, and low-cost but lose performance in cold climates. Flat plate collectors work well for residential, domestic hot water, and solar pool heating. Glazed flat plate versions can reach 140°F for heating applications.

Evacuated tube solar collectors contain rows of glass tubes with the air removed to minimize convection losses. Each tube has a circular absorber fin attached to a heat pipe that transfers heat into a shared manifold. This enclosed design provides excellent heat transfer and hot water output up to 270°F. Evacuated tubes are more efficient in winter and cold climates and more expensive. They work well for commercial hot water and solar air heating.

Parabolic trough-concentrating collectors are long U-shaped mirrors that focus sunlight on an absorber tube running along the focal line of the trough. The tube heats the fluid to 750°F for driving turbines or high-temperature applications. These provide the highest temperatures but require solar tracking. Parabolic troughs are used for large-scale solar power generation, industrial process heat, and solar air conditioning.

Solar dish concentrators are parabolic dish-shaped reflectors concentrating sunlight on a receiver at the focal point. High-intensity heat up to 1500°F can power Stirling engines for electricity generation or provide industrial process temperatures. Dishes require dual-axis tracking but provide the highest temperatures. They are mostly used in large commercial solar farms, ideal for arid climates.

Solar HVAC System Types

Solar thermal HVAC systems come in two main classes: active and passive. Active systems use pumps and controls to regulate solar heat collection and distribution. Passive designs rely on natural convection and architectural elements to trap solar heat. There are four common types of active solar HVAC systems: direct circulation, indirect circulation, isolated gain, and sunspaces/greenhouses. Passive techniques include orientation, window glazing, thermal mass walls, and ventilation. The optimal approach depends on building type, climate, and owner preferences.

Active direct circulation solar designs use pumps to circulate water or fluid directly from the solar collectors to heat exchangers in the HVAC ductwork. This heats the building's air supply directly. While simple, heat exchanger corrosion can be an issue. Direct systems are common in residential hot water and hydronic heating applications.

Active indirect circulation separates the solar and building loop via a heat exchanger. The solar loop circulates between the collectors and heat exchanger with a glycol antifreeze mix that never enters the living space. Indirect designs are more complex but avoid water treatment issues and direct exposure to collectors.

Isolated gain active systems use solar collectors as stand-alone heating devices for ventilation air or hot water. The solar loop and building HVAC system are separate. This approach is easier to retrofit but cannot provide whole-building heating alone. It works well to preheat air or water.

Active sunspaces and greenhouses use glazed solariums and glass rooms as large solar collectors to trap heat. Fans then distribute the warmth to the rest of the house. While aesthetic, active solar rooms lose heat at night and are expensive to construct. They require integration with other HVAC systems.

Passive solar techniques include optimal building orientation, window glazing, and thermal mass walls and floors to capture solar energy naturally. Orienting windows and patios toward the equator maximizes winter solar gain. Strategic overhangs minimize summer solar heat. Passive solar is simple and cost-effective but depends on architectural design.

Solar HVAC Applications and Case Studies

Solar thermal HVAC solutions are a viable renewal energy option for many building types today. Applications include domestic hot water heating, solar combi systems, solar absorption cooling, pool heating, solar air conditioning, and solar ventilation preheat. Commercial and government buildings like offices, schools, military bases, and warehouses are especially well-suited for solar HVAC based on their heating, cooling, and hot water demands. Implementation case studies provide real-world data on costs, performance, and environmental benefits.

Solar domestic hot water heating is one of the most common residential uses. A Florida retirement community home installed a closed loop glycol flat plate solar collector system to preheat water for its electric tankless water heater. The 20-collector array provides 90% of the home's hot water needs, saving $440 annually. With state rebates, the payback period was under 5 years.

Solar systems provide space heating and domestic hot water from the same solar array. A hospital in Colorado installed 96 evacuated tube collectors that feed into a large storage tank. This solar reservoir provides hot water for sinks, laundry, and most building heating. The system saves 2500 MMBTUs and $30,000 in annual energy costs with a 7-year payback.

Solar absorption cooling uses solar thermal energy to drive an absorption chiller for air conditioning. The 50,000-square-foot Roswell Museum in New Mexico uses parabolic trough concentrators to provide solar cooling for exhibit spaces. Solar chillers eliminated $12,000 in annual electric costs with a 9-year payback after incentives. This showcases solar's potential beyond just heating applications.

Efficiency Redefined: Variable Refrigerant Flow (VRF) HVAC Systems

Variable refrigerant flow (VRF) HVAC systems are a highly advanced form of heating, ventilation, and air conditioning for commercial buildings. VRF systems utilize variable capacity compressors, multiple indoor evaporators, and refrigerant flow control to match heating and cooling loads in each zone precisely. This enables exceptional efficiency under partial load conditions, which comprise over 80% of runtime in typical buildings. VRF technology originated in Asia but is now gaining popularity globally due to its flexibility, performance, and energy savings potential.

VRF systems work on the familiar vapor-compression refrigeration cycle like conventional HVAC. However, the key difference is the ability to independently vary the refrigerant flow to each indoor unit based on its real-time demand. This maximizes efficiency by eliminating overcooling and reheating. The variable components, piping, and heat recovery enhancements in VRF systems deliver extraordinary part load efficiency. VRF adoption is accelerating as engineers specify it over outdated central plant designs.

Key Components and Operation

VRF systems have three main components: the outdoor variable refrigerant condensing unit, multiple indoor evaporator fan coil units, and refrigerant piping that connects them. The outdoor condensing unit contains an inverter-driven variable capacity compressor, condenser coil, and expansion valves. It can modulate from 5-100% capacity to match the total building demand.

The indoor terminal units are fan coil evaporators for cooling and heating. Each has an electronic expansion valve that independently regulates refrigerant flow based on its zone setpoint. Network controls coordinate the entire VRF system's variable capacities and flows. Copper piping distributes refrigerant between the condenser and evaporators. Heat recovery units can reuse excess heat or cooling between zones.

In cooling mode, the compressor varies its capacity based on the sum of indoor unit loads. Refrigerant is pumped to indoor unit evaporator coils, which absorb heat and change to low-pressure vapor. The vapor flows back to the condensing unit, releasing heat and condensing to a liquid. Electronic expansion valves at each indoor unit regulate refrigerant flow to match its cooling demand precisely.

In heating mode, the process is reversed. The compressor pressurizes hot gas refrigerant, throttled by the indoor unit expansion valves, and sent to the indoor coil evaporators. As the refrigerant absorbs heat from the air and boils, it provides heating to the zone. The part load capacity modulation ensures that only the required refrigerant is distributed to each coil.

VRF systems can simultaneously heat some zones while cooling others as needed. Excess heat from cooling one space can be redirected to another zone needing heating via heat recovery control valves. This heat recovery minimizes compressor operation for higher efficiency.

The ability to independently control each indoor unit's refrigerant pressures, temperatures, and flow rates defines VRF technology. This provides an incredibly granular demand response to match the exact load in each zone. VRF systems maintain excellent efficiency even at 20-40% of part load conditions, unlike central plant equipment that operates least efficiently at partial loads.

Applications and Usage Examples

VRF systems are well-suited to commercial buildings that require year-round air conditioning and zoning flexibility. The most common VRF applications include:

- Office buildings: VRF allows each office or conference room to have its indoor unit sized for its unique occupancy and loads. Conference rooms can be cooled during events while offices remain off.

- Retail stores: Individual evaporator units can be placed in showrooms, stock rooms, vestibules, and registers to tailor the environment. Units automatically turn off when not needed.

- Hotels: Guest rooms can have dedicated indoor units rather than being tied to a central chiller. Units automatically enter a low-power mode when rooms are vacant.

- Healthcare facilities: Patient rooms, labs, imaging rooms, and offices can all have independent VRF units sized appropriately. Critical environments remain comfortable 24/7 while saving energy.
- Multifamily complexes: Condos and apartments can have one or more indoor units per living space for personalized comfort and efficiency. Units turn off when apartments are empty.

For example, a new 125,000-square-foot office building uses a VRF system with 230 indoor fan coil units ranging from 9,000 to 18,000 BTU capacities. The building has an Energy Star score of 95 and is 40% more efficient than the code minimum. VRF enables right-sized units for each space versus oversized central plant equipment.

Benefits, Savings Potential, and Challenges

VRF HVAC systems provide many performance and efficiency benefits, including:
- Precise load matching and temperature control in each zone
- Simultaneous heating and cooling across different zones
- Full capacity range modulation from 5-100% based on demand
- Reduced ductwork, piping, and space requirements
- Lower energy consumption by minimizing reheat and overcooling
- Heat recovery capability worsens efficiency by up to 30%
- Quiet, distributed operation from multiple smaller units

Studies show VRF systems can achieve 15-30% in energy savings compared to conventional variable air volume systems. The precise load response drives most of these savings. Additional fan power reductions are possible with lower static duct pressures.

While upfront costs for VRF equipment are 20-30% higher, the total life cycle savings lead to a strong return on investment of less than 5 years in many buildings. Rebates, tax credits, and energy incentives from utilities can offset the initial capital costs.

The main challenges for VRF adoption include higher upfront costs, split incentives between builders and owners, lack of familiarity by U.S. designers, and limited technical expertise for maintenance and repairs. However, these barriers are lessening as VRF becomes more popular and technology advances.

FRESH AIR SOLUTIONS: ENERGY RECOVERY VENTILATION (ERV)

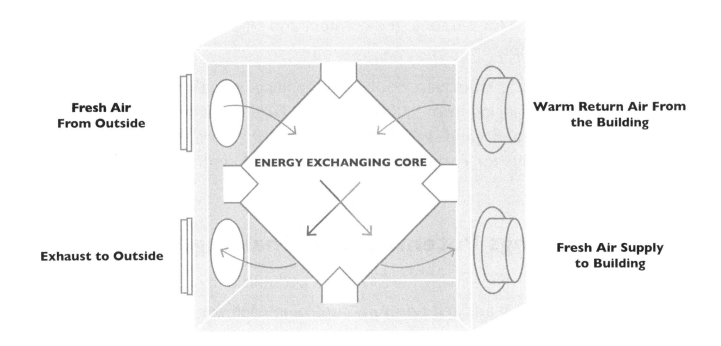

Energy recovery ventilators (ERVs) provide an efficient means of continuously ventilating buildings with fresh outdoor air while minimizing energy losses. By transferring thermal and moisture energy between exhausting and incoming air streams, ERVs can reduce ventilation heating and cooling costs by 50-90%. This makes them a critical tool for sustainably providing healthy, comfortable indoor environments.

How ERVs Work

All buildings require adequate ventilation for indoor environmental quality and occupant health. However, constant intake of unconditioned outdoor air places an energy penalty on HVAC systems. ERVs solve this dilemma through an innovative air-to-air energy transfer process called enthalpy recovery.

The Enthalpy Exchange Process

An ERV core contains very thin passages that allow airflow on both sides while preventing the air streams from directly mixing. As exhaust air leaves the building through one side of the core, its thermal and moisture properties are conducted across the core material to the incoming outside air stream. This pre-conditions fresh air by recovering up to 75% of the "waste" heat and moisture otherwise lost.

Heat and Moisture Transfer

In winter, warm exhaust air gives up its heat to the cold, fresh air stream traveling through the ERV core. This pre-heats fresh air by 10-30°F, reducing the load on furnaces and boilers. During humid summer, moisture is transferred from the cooler, drier exhaust air to the warmer, more humid fresh air. This pre-cools and dehumidifies fresh air entering the building.

Types of ERV Cores

The three main types are plate, matrix, and enthalpy wheel cores. Plate cores contain long, thin, corrugated plates layered closely together. Matrix cores use honeycomb patterns, while wheels incorporate a continuously or periodically rotating desiccant storage media between the air paths. All maximize surface area contact for efficient thermal and moisture exchange.

Compared to HRVs

Heat recovery ventilators (HRVs) transfer heat but not moisture energy between air streams. This makes HRVs less suitable for humid climates where latent capacity is needed. ERVs provide enhanced indoor environment quality and energy efficiency year-round by fully recovering both sensible and latent components of ventilation air.

Components and Operation

Beyond the core, key ERV components work together to ventilate while recovering energy 24/7/365 continuously.

Filters

MERV 7-13 filters in both air streams trap dust and particles down to 0.3 microns for purification before the core. This protects delicate core passages from clogging.

Fans

Forward-curved plug or vane axial fans precisely balance pressure drops through the ERV and building ventilation ducts. Variable speed controls modulate flow rates.

Controls

An integrated electrical panel houses temperature/humidity sensors, programmable logic controllers, timers, differential air pressure gauges, and status LEDs. Configuration settings balance airflows and monitor performance.

Housings

Heavy-duty galvanized steel or aluminum cabinets contain the core, fans, and controls. Insulated compartments separate the two airstreams to prevent crosstalk. Gasketed access doors ease maintenance.

Balanced Flow Design

Careful sizing matches ERV fan capacities to the building's specified fresh air changes per hour (ACH) at a given pressure. This balanced flow prevents back-drafting or pressure imbalances that waste energy or impair indoor quality.

Proper Installation

ERVs are mounted indoors within 10 feet of the building envelope for the shortest duct runs. Intake and exhaust connections tie into the ventilation system. Grade-level units require protective shelter from the weather. Start-up commissioning ensures airflow balancing and effective operation.

ERV Efficiency & Benefits

In the real world, studies repeatedly confirm ERVs fulfill their energy-saving promise. Their advantages span improved comfort, air quality, humidity control, and reduced operational costs.

Energy Savings

Field tests show ERVs cut ventilation heating energy use by 70-90%. Cooling demand decreases by 50-60% by pre-conditioning fresh air all year. The U.S. Department of Energy estimates typical payback periods from utility bill savings only of 2-5 years.

Heating Season Dominance

Winter months require the largest space conditioning energy input to treat ventilation loads. Pre-warming fresh air with an ERV is exponentially more economical than reheating it with furnaces or boilers with higher fuel costs and lower efficiencies.

Summer Benefits Too

While cooling savings may seem less dramatic, dehumidifying and pre-cooling fresh air still prevents added humidity and thermal stress on the air conditioner. Reduced condensation also hinders mold growth. HVAC equipment runs less frequently under lighter total ventilation loads.

Air Quality Assurance

ERVs protect indoor environments from pollutants by mechanically pressuring the building envelope to purge moisture, odors, and combustion byproducts. Automatic control ensures code minimum fresh air exchange rates are consistently maintained.

Superior Humidity Management

At indoor relative humidity (RH) settings of 30-60%, ERVs transfer moisture between airstreams to stable comfort levels without wasting energy on over-conditioning. These low RH conditions inhibit dust mite growth and other triggers for allergies and respiratory issues.

Comfortable Transition

Gradual pre-tempering of ventilation air through the core means occupants perceive no drafts upon even large open door-to-outdoor temperature differences, promoting well-being.

Noise Modeling

Sounds within occupied spaces are unaffected. ANSI-rated ERV sound power levels produce ≤NR30 interior noise levels that meet strict codes. Locating units as far as possible from rooms minimizes inaudible background sounds.

Low Maintenance

Well-engineered filtration and rotating/non-rotating core designs need only periodic filter changes and sensor/controller calibration to ensure continuing performance. Most components last 15+ years, far surpassing simple payback periods.

Healthy IEQ For All

Proper use of ERV technology effectively distributes fresh, clean air throughout buildings equitably while sustaining efficiency. This impacts productivity, attendance rates, reduced sick days, and medical expenses - benefits impossible to assign a monetary value. Quality indoor air is a basic necessity enabling occupant wellness.

ERV Applications & Sizing

Suitable applications include schools, healthcare and laboratory facilities, clean rooms, offices, multifamily structures, commercial kitchens, and more - essentially any building category subject to ASHRAE, IMC, or local ventilation code requirements. Proper sizing is critical to achieve latent and sensible recovery potential.

Ventilation Load Determination

Using building plans and occupancy schedules, design engineers calculate fresh air changes per hour (ACH) and total cfm (cubic feet per minute) of outdoor air to treat based on ASHRAE Standard 62.1 or local mechanical code.

ERV Size Selection

Manufacturers publish selection guides for specific models based on intake/exhaust flow rates and internal pressure drops. Published performance tables provide recovery effectiveness (%), static pressure specifications, and utility connection requirements. Over- or under-sized units fail to meet balanced flow or energy recovery goals.

Variable Flow Options

Certain ERV products offer variable speed controls, inlet dampers, or integrated VFD drives to modulate flow rates spanning 10-100% of rated capacity. This allows balancing to actual rather than estimated loads and fine-tuning flow balance based on blower door or duct blaster tests.

Placement & Duct Work

Indoor wall or curb mounts adjacent to the building envelope minimize duct runs. Intake and exhaust ducts route to/from the ERV and connect with main ventilation system branches. Correct insulation helps prevent condensation.

Commissioning

A testing, adjusting, and balancing (TAB) firm verifies airflows match design and balance inputs/outputs within recommended tolerances. Controls functionality, sensor calibration, and troubleshooting deficiencies before occupation ensures optimized building performance.

Application Case Study

Consider a new 50,000 sq. ft. elementary school with an expected population of 400 students and staff during operating hours. Using ASHRAE Standard 62.1 ventilation rate guidelines:

- Occupancy of 2 people/1000 sq. ft. yields 100 occupants
- Outdoor air required = 5 cfm/person = 100 * 5 = 500 cfm

- With 100% fresh air and no airside economizer, the ERV must provide 500 cfm of ventilation air.

Based on manufacturer selection software:

- A 660 cfm plate core ERV model rated for 80% maximum total energy recovery effectiveness meets the 500 cfm load with safe pressure and static specifications.
- Unit installation includes mixing plenum duct connections 10 feet from the building wall.
- Third-party TAB confirms balanced 500 cfm flows within tolerances.
- Energy modeling shows annual ventilation heating gas usage reduced by 80% versus no ERV. Simple payback estimated at 3 years.

Proper ERV sizing and installation ensures this school provides ongoing healthy indoor environments efficiently for decades. The unit easily pays for itself multiple times over through lower operational costs.

FUTURE TRENDS IN HVAC: WHAT LIES AHEAD

Looking towards the coming decades, several prominent innovative trends are poised to transform the HVAC industry:

Artificial Intelligence and Machine Learning

Deep learning algorithms will optimize entire HVAC systems for maximum efficiency based on big datasets encompassing external and internal conditions, occupancy patterns, weather forecasts, and more. Utilizing AI geothermal balancing, dynamic refrigerant controls, IoT integration, and predictive diagnostics could surpass human abilities.

- **AI System Management:** Self-monitoring "digital twin" models will detect performance anomalies, predict failures, and automate repairs before issues arise. AI will learn unique facility characteristics and tailor settings accordingly through continual adaption.
- **Intelligent Controls:** Rather than programmed sequences, AI controllers will make autonomous second-by-second decisions, factoring in numerous variables. Coupled with IoT integration, AI systems will precondition spaces before occupancy based on predictive algorithms.

- **Machine Maintenance:** Robotic endpoints, computer vision, immersive VR/AR, and exoskeleton aids will augment human abilities for remote inspections, troubleshooting complex areas, and hazardous maintenance tasks from any location. AI assistants will guide non-experts.

- **Personalized Comfort:** Occupant preferences, habits, and biological profiles will define individualized 'comfort zones.' AI will provide personalized heating, cooling, and IAQ perfectly optimized for each user or space type based on deep-learned patterns.

Connected Buildings

Hyperconnectivity and two-way communication will allow integrated monitoring, control, and feedback of all building systems via the Internet of Things (IoT). Pervasive sensor data and cloud analytics will transform facility management.

- **Remote Access:** From smartphones, augmented reality interfaces, or smart home assistants, authorized personnel will monitor live equipment health/performance, automate schedules, receive alerts/alarms, and remotely troubleshoot any issues.

- **Sensor Networks:** Embedded environmental, air quality, occupancy, and smart metering sensors will transmit real-time statistics on conditions, energy usage, and more to central control platforms for analytics and informed decision-making.

- **Predictive Maintenance:** Sensor fusion with AI patterns will enable condition-based maintenance by detecting subtle performance degradations early. Resource allocation will focus on preventing critical outages or failures that waste energy and money.

- **System Health:** Connectivity allows continuous remote verification that systems operate as designed and specifications are met. Automated notifications of deviations, malfunctions, or suboptimal control sequences streamline corrective actions.

- **Big Data** Applications: Facility managers will harness sensor data mining, analytics, and reporting to improve building/equipment operations based on key performance indications. Quantified results incentivize sustainability improvements for measurable reductions in carbon footprint and costs.

Renewable Integration

Renewable energy integration with HVAC systems will lead to decarbonizing buildings. Creative hybrid solutions capitalizing on local climate resources will emerge.

- Solar Thermal Absorption Chillers: Parabolic trough collectors or solar thermal panels heat water in absorption chillers to produce chilled water for cooling. Thermal storage buffers intermittencies for peak AC loads.

- Geothermal-Solar: Borehole thermal energy storage coupled with above/below ground electro-geothermal systems provide limitless hot/chilled water for hybrid heat pumps. Complemented by PV for on-site renewable electrification.

- Community Microgrids: District renewable energy systems leverage proximity benefits to provide zero-emission HVAC using synergistic combinations of solar, wind, geothermal, hydropower, and seasonal thermal storage. Utility buy-back programs incentivize surplus renewable exports.

- Smart Hybrid Controls: Integrated IoT-connected control platforms optimize cycling between multiple renewable energy sources, including thermal, photovoltaic, wind, and stationary batteries, based on fluctuations in availability and demand. AI self-learning components maximize renewable HVAC fraction.

- Renewable Nanogrid Microclimates: At building-scale, hybrid PV-battery-fuel cell-heat pump nanogrids condition interior environments autonomously using on-site renewables. Zero reliance on fossil-fuel infrastructure establishes sustainable, climate-resilient, electric microclimates.

Grid Harmonization

As renewable penetrations increase nationwide, intermittent clean power sources necessitate new approaches to coordinating generation and loads. HVAC loads represent a promising dispatchable flexibility resource.

- **Demand Response Programs:** Utilities may incentivize shifting HVAC operation to off-peak via time-of-use rates or reducing loads on command. Programmable WiFi thermostats automatically participate anonymously in demand reductions that balance variable renewables.

- **Thermal Energy Storage:** Integrated thermal, chilled water or ice storage allows shifting HVAC loads to peak renewable output hours, leveraging inexpensive excess wind/solar. Discharging storage maintains comfort despite supply intermittencies.

- **Managed Distributed Energy Resources:** Coordinated controls dispatch controllable distributed energy assets like HVAC in microgrids and campuses as a virtual power plant. DR programs value flexibility resources, providing grid services supplementing or deferring large generation and transmission upgrades.

- **Integrated Community Energy:** District energy systems incorporating centralized CHP, geo-exchange, renewable heat recovery, and infrastructure-sharing optimize renewable energy usage across multiple buildings. Centralized controls harmonize aggregated loads and resources for maximum community efficiency.

Carbon-Free Future

Accumulating innovations will drive HVAC towards a future decarbonized entirely through renewable options. While challenges remain, determined effort and judicious policies can accelerate progress to eliminate fossil fuel reliance and achieve a net-zero balance eventually. HVAC systems will be driving through optimized, integrated designs that reduce environmental impact. With further advances in renewable and storage technologies, a carbon-free future is well within reach.

PART 4
COST REDUCTION & REAL-LIFE EXAMPLES
PART 4

CHAPTER 7
Tips for Immediate Cost Reduction

Maintaining an HVAC system efficiently across the heating and cooling seasons can help significantly reduce long-term energy costs. However, homeowners often want fast solutions that deliver savings immediately. This chapter provides tips and strategies readers can immediately implement to cut their HVAC expenses. You'll start seeing a difference on your next energy bill with some minor adjustments or simple DIY projects. Let's explore our options for quick wins.

Making small behavioral or maintenance changes around the house can lower utility costs in a very short period. This expanded section provides greater details on impactful quick fixes homeowners can try.

Replace Air Filters Regularly

Replacing HVAC air filters monthly is one of the easiest and most impactful things you can do to reduce energy costs immediately. A dirty filter blocks airflow into the HVAC system and forces the blower and air handler to work much harder to push air through the restricted filter. When the system exerts more effort to move the same amount of air, it becomes less energy efficient. It uses more electricity to power the blower and compressor at a higher output continuously.

An efficiency-rated HVAC system is designed to perform optimally when household air flows through regularly maintained and cleaned filters. Dirty filters block airflow and disrupt this intended design, like putting your car engine under extra strain by forcing it to push air through a clogged air intake. Over time, this decreases the overall efficiency and capacity of the HVAC system.

The Department of Energy estimates that regularly replacing or cleaning furnace filters can improve a home's heating and cooling efficiency by anywhere from 5% to 15%. This means potential savings of hundreds to over a thousand dollars yearly, depending on the size of your home and the HVAC system. Replacing dirty filters with new ones is like pushing the reset button that allows your HVAC system to run as efficiently as when it was first installed.

Most experts recommend replacing filters every 30 days if you have pets or 90 days for clean households. Sliding the new filter into the supply or return vent takes under a minute but immediately keeps your home comfortable while lowering cooling and heating costs. Proper disposal of used filters prevents dust and debris from recirculating into the living areas. Replacing filters once a month delivers fast and lasting savings with minimal effort.

Clean or Replace the Furnace Vent

Over the years, furnace exhaust vent pipes outside the home naturally accumulate dust, dirt, pollen grains, and residues from continually expelling combustion gases.

This buildup inside metal vent interiors forms restrictions that obstruct smooth airflow. Like clogged arteries, restricted vent pipes increase back pressure in the furnace, forcing it to work harder to push exhaust out. The furnace then uses more fuel to maintain the same heat output levels.

Vent pipe clogs silently but significantly lowers combustion efficiency. To clean, homeowners can use a telescoping brush handle and wire brush attachment specifically made for furnace vents. Working from the rooftop vent or side wall termination, gently brush the interior surfaces from the furnace end outward. Remove built-up obstructions to restore vent diameter fully. Wear eye protection since particles dislodge during cleaning.

For severe blockages, replacing sections of corroded or molded venting may be necessary. Home centers stock bent steel pipe fittings designed for do-it-yourself installation. Cut out restricted segments with a reciprocating saw and join new sections according to the configuration code. Investing an hour's work to clean or swap vents notices an instant drop in gas usage that benefits your budget and the environment. When vents flow freely again, the furnace can burn fuel more cleanly and efficiently for life-extending performance.

Check Ductwork for Leaks

After the furnace generates warm or cool air, it relies on an interconnected duct system to properly distribute conditioned air to living spaces. Over the decades, ductwork insulation and seams can crack or weep air leaks from general wear and natural vibrations inside walls. Like furnace vents, duct leaks represent a path of least resistance that wasted conditioned air escapes through before circulating through the home. This results in higher HVAC runtime continually trying to reheat or recool otherwise lost air.

Homeowners can conduct a basic inspection of ductwork for indications of air leaks. Start by switching on the HVAC blower fan and feel the supply runs with the back of your hand. Leaks feel drafty, while isolated areas may kick up dust. Visual signs include musty odors, fiberglass duct insulation detaching, or puffs of dust emerging from cracks. Leak locations are often in attic spaces near air handler connection points, behind access panels, or junctions in the basement or crawlspace.

Once pinpointed, seal openings with latex mastic, aluminum tape, or butyl rubber tape products made for duct sealing. These adhesive sealing treatments adhere permanently to prevent conditioned air escape even under system pressure changes. Stopping duct air leaks is cost-effective since it retains heating/cooling energy inside the home structure. Within a month or two of eliminating leakage sites, energy bills will clearly show savings from improved duct system efficiency.

Insulate Exposed Ducts and Pipes

Any section of ductwork or water pipes leading to and from the HVAC system located in unheated or uncooled spaces like crawlspaces, attics, unfinished basements, or garages quickly loses a significant amount of thermal energy if not properly insulated. Outside air directly interacts with uninsulated metal surfaces, readily stealing heat or coolness before spreading through the home. This forces the HVAC system to work to replace lost thermal capacity, wasting energy continuously.

A common culprit area is the final 5-10 feet of flex duct leading between the air handler unit located in an attic and the supply vent register outlets in the ceilings below. Heat rises through an unconditioned attic, rapidly drawing out capacity from duct surfaces. Likewise, non-insulated pipes carrying hot water for heating systems or chilled water for cooling setups disposed of through less thermally controlled areas are prone to heat loss or gain.

To prevent energy bleeding, wrap all exposed duct sections and pipes with fiberglass insulation sleeves rated for exterior temperatures up to 250°F. Secure insulation with staples, metal wire, or duct tape. For ducts, use R-8-rated insulation for at least 1-inch thickness. Pipes call for a 1.5-inch R-12 minimum. Insulation acts as a thermal barrier, stopping temperature transfer between the air inside and surrounding environmental temperatures outside. This retains capacity within the distribution system more efficiently to heat or cool living spaces without wasted effort by HVAC equipment. With a small material investment, insulation pays dividends through energy conservation immediately.

Adjust Thermostat Settings

The HVAC thermostat control sits at the heart of any home comfort system, regulating operations for climate control based on set temperature thresholds. Minor temporary tweaks to programmed thermostat settings offer relatively fast savings returns through decreased HVAC runtime. While customized for residents' lifestyles, a few degree adjustments when home versus away frequently go unnoticed but save substantially on monthly utility bills.

For cooling costs during spring and summer, raising the thermostat setpoint by 2-3°F when the home is a simple change activates less demand on the air conditioner. At 78°F rather than 75°F, indoor conditions still feel mild enough for most households yet cut related energy usage by approximately 5-15%, according to the Department of Energy. Savings appear right away through reduced A/C runtime. Overnight or when away at work, even half a day, increasing the set temp a further 3-5°F lessens the cooling workload without complaints.

Likewise, lowering the heating thermostat setpoint a few degrees from the usual 72-75°F range offers quick savings during winter. Dropping to 70°F when home delivers 10-20% heating fuel reductions immediately. When away or sleeping, reducing heat another 3-5°F allows the furnace to cycle off for longer periods with minimal sacrifice to indoor comfort levels upon returning. Programmable or smart WiFi thermostats make implementing custom multi-stage temperature schedules easy for maximum convenience and budget benefits.

HVAC energy usage decreases noticeably by simply adjusting existing thermostat settings more thoughtfully on a temporary basis according to daily routines. Families help cut monthly bills promptly without spending money on upgrades through smarter control of their most powerful efficiency tool – the thermostat. Fine-tuning temperature thresholds delivers fast returns while satisfying comfort needs appropriately.

Lower Window Treatments

During winter heating and summer cooling seasons, passive solar heat exchange plays a meaningful role in indoor temperatures that HVAC systems then manage. On sunny winter days especially, the sun beams copious free thermal energy through windows, gradually warming interior surfaces and air if allowed contact. For each degree the sun elevates indoor conditions naturally, the furnace saves fuel by not having to generate that heat artificially.

Opening curtains, blinds, shades, or drapes on east, south, and west-facing windows utilizes solar preheating effectively from dawn until dusk. Interior temperatures slowly rise throughout daylight as sunlight permeates furniture, walls, and floors with warmth holdable long after sunset. Homeowners can save 5-10% annually on winter fuel bills by maximizing passive solar collection in this simple manner. Close coverings again at nightfall to maintain absorbed heat indoors overnight.

Conversely, keeping all glass exposures shaded from direct sunrays becomes more important during summer. Blocking solar gain through windows prevents interior spaces from overheating, decreasing the load on air conditioners. Allow sunshine access in cooler months for comfort and savings. Temporarily adjust coverings each morning and evening based on the season for immediate impact from "solar-powered" climate control. With no outlay, optimized window treatments help reduce bills promptly.

Lubricate Fan Motors

Like any machine with moving parts, HVAC fan motors transporting indoor air gradually experience internal friction that raises operating costs if left unaddressed. Dust and lint particles accumulate over 5-10 years, bearing grease and causing motors to work harder to spin fans smoothly. This increases energy draw and prematurely shortens component life expectancies.

Disconnecting electricity and thoroughly cleaning away residue builds within fan housings restores like-new functionality. Dilute degreaser sprayed onto cotton rags or Q-tips reaches all surfaces and edges. Take time to unclog any dust-packing motor shaft openings or edges that impede free-spinning.

Reapply lightweight lubricating oil sparingly to lubricated bearing surfaces with an applicator before reinstalling protective fan blade guards. Motors churning freely on oiled shafts and bushings reduce energy costs through lower rotational resistance. Regular lubrication delivers prompt and ongoing savings in utility bills besides prolonging expensive component replacements later.

DIY PROJECTS FOR INSTANT SAVINGS

Larger HVAC maintenance tasks are also relatively simple for homeowners with basic skills. The following fast DIY fixes deliver immediate returns without a service pro.

Clean or Replace Furnace Burners

Gas furnaces rely on burners inside the firebox chamber to generate heat by mixing gas fuel with air and igniting a controlled combustion flame. Over many heating seasons, burners collect thin films of residues and deposits from dust, lint, and combustion byproducts pulled into the furnace system through return ducts.

While a light accumulation may not impact performance significantly, heavy buildup interferes with efficient heat transfer.

Residues form an insulating barrier on burner surfaces that prevents full heat transfer to surrounding metal components. The furnace must compensate by burning higher gas volumes to reach the same internal temperature levels. Over time, this inefficient operation increases heating costs due to increased fuel consumption. To maintain optimal efficiency, burners should be cleaned or replaced every 5-10 years, depending on the home's air cleanliness.

Inspecting burners involves removing the furnace access panel and carefully examining each burner body and port using a flashlight. Intact burners with only a thin coating may be gently cleaned using a soft wire brush or vacuum attachment. However, heavy sooting, cracks, warping, or other signs of deterioration indicate replacement is necessary. Homeowners can purchase new ceramic burners online or at home supply stores to match their furnace make and model.

Replacing burners takes only an hour with basic hand tools. First, turn off the power to the furnace and gas line before removing any wires or tubing attached to the old burners. Take photos for reference if needed. Carefully pull out damaged burners and replace them with new ones, properly lining up gas and ignition ports. Reattach any connections and replace access panels before turning the systems back on. With like-new burners heating efficiently, fuel bills see immediate relief.

Change Thermostat Batteries

Proper thermostat function relies on consistent battery power to regulate HVAC systems as intended. Weak or dying batteries disrupt temperature control and waste energy to recover setpoints during fluctuations. Most digital thermostats use common AAA or AA alkaline batteries, typically providing around one year of runtime. However, programming and sensors still drain power slowly, even without frequent use.

Many models sound low battery warnings as levels deplete, but it's best to replace thermostat batteries twice yearly for routine maintenance. This ensures reliable, precise operation throughout each season. Replacement only takes a few minutes with basic tools. First, locate the small screw or switch on the rear or base to release the front cover/faceplate. Slide out the old batteries and dispose of them properly, then insert new ones. Pay attention to polarity markings to avoid damage.

Reattach the cover and confirm the thermostat returns to normal functioning. Test all settings and features to be certain. Regular battery swaps can prevent frustrating malfunctions from heating/cooling ineffectively or accidentally changing programmed schedules. Proper power keeps HVAC systems running at their comfortable, energy-efficient best around the clock with minimal effort.

Clean Condensate Drains

Modern air conditioners and humidifiers produce condensate water collected from the air during dehumidifying. This condensate must be safely drained from the appliance through plastic pipes to a suitable outdoor location like a floor drain or back patio. Unfortunately, mold, algae, or sediment buildup that blocks proper drainage can obstruct pipes over time.

Clogged condensate lines cause water backflow issues, corrupted air circulation, ineffective dehumidifying, and potentially expensive water damage if overflow occurs. However, cleaning the drainpipes takes just 30 minutes and restores optimal performance. First, disconnect the power to the unit and locate the main condensate line originating underneath or behind it. Remove any visible caps or cleanout plugs to access the pipe interior.

Inspect the line for signs of blockage like bulging or standing water. Use a nylon bottlebrush, coat hanger, or drain snake bent into an L-shape to remove debris clogging the pipe from the appliance outward. Flush with white vinegar and water if needed to dissolve the film. Reinstall cleanout caps and reconnect power once water flows freely again. With this fast and simple seasonal task, watch for improved dehumidifying and protection against future drain-related breakdowns.

Inspect Heat Exchanger

The furnace heat exchanger cavity constantly undergoes thermal expansion and contraction during the heating cycle, which can gradually cause cracks in less durable materials over 10-15 years. Cracks compromise heat transfer and gas containment, creating unsafe, inefficient operations. While replacement may ultimately be necessary, basic annual inspection allows homeowners to catch small issues early.

Remove the burners and view the exchanger through the opening using a flashlight. Check for hairline cracks along seams or joints in metal constructions. Also, look for colored surfaces indicating potential corrosion developing internally. Caressing fingers between exchanger fins makes for gaps or brittle sections. Continued use may be safe with professional guidance if cracks appear minor with no pouring exhaust or modeled coatings.

However, severe cracking or breakdown signs require a qualified technician examination and potential part replacement before operating further. Safety overrides minor repair costs, so erring on the cautious side preserves life and property. Regular exchanger checkups extend furnace lifespan by nipping expensive damage in the bud with quick scans.

NEGOTIATING WITH HVAC CONTRACTORS: INSIDER TIPS

Hiring professional help for major HVAC replacements, repairs, or installations often involves weighing service quality against contractor rates. While several quotes seem ideal, this approach risks confusion or hard sales better avoided through informed preparation.

Research Average Local Rates

Review websites and HVAC association materials stating common installation or repair costs to determine market norms. Know fair pricing upfront rather than overpaying. Request complete itemized invoices for transparency on labor and parts. Reputable contractors welcome informed customers.

Request Service Plan Details

Prepare proposed maintenance package specifics, including covered checks, response times, parts/labor warranties, and total projected annual costs. Request contractor standard plans for direct comparison in negotiations.

Ask for Referral Contact Information

Request 3-5 reference names, addresses, and phone numbers from past satisfied customers. Call references to discuss work quality, professionalism, and overall service experiences before deciding. Positive reviews indicate trustworthy technicians.

Inquire About Financing Options

0% interest payment plans let large projects proceed affordably. Factors like flexible 6-12 month terms influence choice if quality/experience also aligns. Long assistance pays higher initial investments comfortably.

Bundle Maintenance with Installations

Bundling installation quote with a 3-5 year maintenance contract secures continued priority service.

Negotiate contract discounts such as free annual tune-up inclusion. Committing long-term satisfies contractors and saves customers 10-15% in total project costs.

Get Agreement Details in Writing

Confirm and document all quoted amounts, payment schedules, warranties, projected equipment lifespan milestones, and maintenance/service commitments within the signed proposal.

A written contract protects against future billing discrepancies or void warranties.

Preparation lets homeowners confidently choose reputable companies providing the best-combined service, equipment quality, and pricing structures suited to individual household needs and budgets. Negotiating professionally maximizes value from HVAC professionals.

UNDERSTANDING YOUR ENERGY USAGE: TOOLS AND RESOURCES

A clear picture of household energy consumption patterns empowers homeowners to target the most cost-effective efficiency improvements. Various reporting utilities and online tools offer consumption analytics for evaluation.

Review Past Energy Bills

Evaluate previous 12+ months of utility statements to spot seasonal billing and usage fluctuations. Note monthly highs/lows aligning with temperature trends. Identify if rates charged are competitive and inquire about cheaper energy provider options. Utility statement histories contain rich usage and cost insights.

Analyze Online Energy Portal Data

Major gas, electric, and oil service utilities provide online customer login portals for viewing custom energy usage data. Graphs display daily, weekly, and monthly consumption statistics. Patterns show peak demand periods aligning with temperature highs/lows or daily routine timing. Such real-time insights reveal where tweaks can lower energy costs.

Schedule a Home Energy Audit

Contact the servicing utility to arrange for a professional home energy audit at no or low cost. Auditors perform a field assessment analyzing the home blueprint, insulation/air-sealing, appliance/lighting efficiency, and HVAC system performance. Infrared cameras detect thermal leaks.

Blower doors measure air changes. Auditors subsequently provide a customized report outlining specific improvement recommendations and potential rebates available for implementing auditor-suggested upgrades. Their unbiased expertise helps prioritize projects for maximum savings and returns.

Review Smart Thermostat App Data

Programmable and Wi-Fi-enabled smart thermostats come equipped with free companion smartphone applications displaying system runtime insights. View daily and weekly HVAC activation patterns in graphs.

Abnormal activations outside the schedule indicate possible programming errors or thermostat faults. Advanced unit apps integrate with voice assistants, too, for remote comfort controls. Thermostat analytics aid optimization.

Benchmark Against Similar Homes

Websites like EnergyStar.gov and HomeEnergyScore.gov allow entering a home address, square footage, occupancy details, and utility provider to generate customized benchmarking reports.

Results compare the home's current energy usage to similar efficient homes in the same climate zone to uncover if above or below-average users should investigate improvement potentials. National median scores indicate room for reduction efforts.

Track Individual Circuit Usage

Portable energy usage monitoring devices let homeowners plug major appliances, HVAC components, and other circuits/outlets into the device to view real-time power consumption data.

Isolate energy-guzzling or inefficient "vampire" loads for upgrades. Such sub-metering empowers targeted projects, maximizing limited budgets and effort. Fact-based decisions drive the most savings impact.

CREATING A CUSTOMIZED HVAC MAINTENANCE SCHEDULE

Regular seasonal maintenance extends equipment life expectancies while preserving peak operating efficiency. Developing a customized schedule tailored to individual household needs keeps HVAC systems functioning smoothly all year round.

Spring Startup (Late March-May)

Inspect furnace/air handlers for cracks, clean burners, and heat exchangers. Check for damaged wiring, ducts, or insulated pipes. Examine flue pipes/chimneys, ensuring unblocked exhaust flow.

Test thermostat and humidifier functions. Consider professional service if over 10 years old to avoid potential service calls. Recharge the A/C system if under 5 years to maintain refrigerant levels.

May (Annual)

Replace furnace and AC unit air filters. Clean or vacuum lint screens in the furnace cabinet and condenser coils outside. Tighten electrical connections if corroded. Lubricate the fan motor if accessible.

Examine ductwork for air leaks, sealing with mastic or tape. Remove debris from drain pans and lines to prevent overflows.

Summer (June-August)

Repeat monthly filter changes and clean outdoor condenser coils and fins to remove dust, pollen, and mold buildup, reducing efficiency. Inspect ductwork, thermostat settings, and cool air distribution.

Trim vegetation away from the condenser for proper airflow. Check the roof for leaks potentially damaging equipment underside.

Fall (September-October)

Service humidifiers, test duct balancing, and modify dampers/registers if required. Audit duct and pipe insulation for integrity against the coming winter. Replace filters before the heating season. Clean furnace components again if used minimally over warm months. Test backup heating system functionality before use.

Winter (November-February)

Monitor furnace and structures for potential carbon monoxide leaks requiring professional diagnosis. Keep snow and ice clear away from outdoor condenser units and vents. Check thermostat batteries' useful life. Ensure gas line regulators deliver safe furnace fuel pressures. Consider hard-wired CO detectors for extra safety assurance in high-risk rooms like attached garages.

Expanding a half day annually on proper maintenance safeguards the HVAC system and home while reducing utility costs by up to 15% versus neglected systems requiring costly repairs down the road. Small efforts deliver big benefits.

CHAPTER 8

Case Studies and Success Stories

Real-life examples can be tremendously helpful for demonstrating the tangible benefits of proper HVAC system management and maintenance. This chapter will examine several case studies and success stories that showcase dramatic cost savings, expert strategies, transformations in homes and businesses, and inspirational examples of others who have optimized their HVAC systems. By learning from these real-world experiences, readers can glean practical insights for leveraging.

CASE STUDY: ANNUAL SAVINGS OF $2,000 THROUGH PREVENTATIVE MAINTENANCE

The Johnson family had struggled with unexpected repair bills for their home AC system for years. Fed up with the mounting costs, they decided to try a new approach.

Scheduling Regular Checkups

The first thing the Johnsons did was schedule annual pre-season checkups with their HVAC contractor. During these appointments, technicians thoroughly inspect the outdoor unit, indoor coil, ductwork, filters, thermostat, and other components. Any issues identified during inspections could then be proactively addressed before becoming serious problems. This proved very valuable, as smaller problems that were caught could be fixed inexpensively before causing larger failures.

Replacing Aging Components

One of the checkups revealed that the indoor coil was showing signs of corrosion and needed to be replaced. Since a failing coil could compromise cooling performance and efficiency, the Johnsons upgraded it. Replacing the coil before it failed helped avoid a full system replacement that would have been far more expensive.

Improving Airflow and Filtration

The technicians also noticed that ductwork connections were leaky, and filters were overloaded with debris. Upgrading to a higher-grade air filter and sealing ducts with mastic solved these airflow and filtration issues, improving overall system efficiency. Catching issues like duct leaks allowed them to be fixed at a minimal cost compared to the potential damage they could cause.

Optimizing Settings and Controls

Additionally, the thermostat was programmed with a more efficient schedule, and the indoor fan was set to continuous mode for better air circulation and humidity control when heating or cooling wasn't running. These small control adjustments helped improve comfort and lower energy bills.

Measuring Energy Savings

The Johnsons avoided any unexpected breakdowns during that cooling season by implementing the contractor's preventative maintenance recommendations. Plus, their utility bills were $200 per month lower thanks to efficiency improvements. Over the year, they saved $2,400 with their new proactive approach. The hard numbers proved that maintenance paid off financially.

Renewing the Maintenance Plan

Impressed by the results, the family signed up for ongoing annual maintenance to sustain these savings long-term. The minor investment saved them thousands versus constant repair bills down the road. Continuous maintenance ensured the system remained in good working order and at peak efficiency.

Spreading the Word

Thrilled by their experience, the Johnsons shared it with friends and neighbors. Multiple homeowners they referred also adopted preventative maintenance and saw similar cost reductions. Word-of-mouth is powerful advertising for any HVAC contractor.

Valuable Lesson Learned

This real-life success story demonstrates how a systematic maintenance plan can help avoid pricey surprises while improving efficiency. The Johnson family's case proved that spending a little to service HVAC components pays off exponentially in reduced energy costs and expenditures on fixes later on. Their positive outcome showed others the importance of active system stewardship.

Referrals Yield More Business

The contractor picked up several new maintenance agreements that year based on referrals from satisfied customers like the Johnsons. Good work and customer satisfaction drive continued career success in HVAC.

Preventative Maintenance Pays Dividends

Five years after implementing their maintenance program, the Johnsons continued to see yearly savings of $2,000 or more. Their HVAC equipment remained in top working condition with no unexpected breakdowns. Small upfront costs translated to huge long-term financial and reliability benefits.

John Smith has worked in the HVAC industry for over 30 years as a licensed contractor. Here are some of the maintenance tactics he recommends to clients.

- **Seasonal System Checks:** Like the Johnsons' contractor, John believes the best defense is a good annual inspection before each new heating and cooling season. Catching smaller issues early can prevent major repairs down the line. This approach helps find minor problems when they're cheap to fix.

- **Preventative Replacements:** He also looks for components nearing the end of their lifecycles, especially older blowers, compressors, coils, and thermostats. Replacing them preemptively heads off seasonal breakdowns. John has found equipment lives much longer when proactively maintained this way.

- **Filter Changes:** Ensuring filters are replaced monthly avoids restricted airflow that stresses motors and decreases efficiency. John stocks clientele with bulk filter purchases for consistency and convenience. Clean filters are critical to performance.

- **Thermostat Management:** Optimizing a home's programming, settings, and controls greatly saves energy costs. John reviews each thermostat installation annually and adjusts settings as needed. Proper use offers great savings potential.

- **Humidifier/Dehumidifier Sanitizing:** Mineral deposits clog these units quickly, so an annual vinegar or bleach cleanse restores optimal moisture control without costly replacements. Proper cleaning is neglected maintenance that can cause expensive failures.

- **Refrigerant Inspections:** John most strongly advocates biannual gauging of refrigerant levels, which are vulnerable to leaks. Catching low charges before failure season can prevent a full system overhaul. Leaks are difficult to notice yet can destroy efficiency.

- **Client Education:** Equally key is teaching homeowners about general system operations and signs of issues to watch for. An informed client maintains better and calls promptly if help is needed. Educated customers result in fewer emergencies and better care overall.

- **Proficiency Pays Off:** With diligent adherence to professional practices over three decades, John's business has flourished through word-of-mouth as customers praise his record of reliability and annual negotiated maintenance packages. His strategy sustains systems for maximum lifespans. John serves as a great example of success in HVAC contracting.

- **Focus on Longevity:** John finds that emphasizing long-term value and reliability over low initial costs keeps customers satisfied for life. Quality work and maintenance ensures equipment lasts 15-20 years as designed rather than premature failures.

- **Valuing Trusted Relationships:** He enjoys multi-generational clients recommending family and friends for HVAC needs. John credits his tradition of transparent service for maintaining strong customer bonds over the years. Winning loyalty rewards with ongoing business and referrals.

OVERHAUL SAVES THOUSANDS FOR A SMALL BUSINESS

A local restaurant had endured five summers of inadequate cooling due to an antiquated rooftop package unit nearing 30 years old. This story recounts their air conditioning transformation.

- **Assessing the Aging System:** A professional contractor inspected and found remnants of a long-defunct refrigeration cycle, corroded internals, and inefficient fans and motors consuming excessive power. The unit was well past its prime and struggling to handle the restaurant's load. A comprehensive evaluation revealed it was at the end of its usable life.

- **Recommending a Modern Upgrade:** Given the age and disrepair, replacement emerged as the smartest financial decision versus attempts at repairing the outdated setup. The unit was wasting huge amounts of energy and struggling on hot days. A new system with today's technology would dramatically improve efficiency and comfort.

- **Installing Advanced Components:** Contractors removed the old package and installed a high-performance roof-mounted system utilizing eco-friendly refrigerant. Electronically commutated motors, advanced controls, and onboard diagnostics also enhanced capabilities. The owners invested in a future-proof system for long-term savings and reliability.

- **Noticing the Difference:** Upon startup, owners and staff were amazed by the leap in cooling strength and quieter operation. Unlike frustrating past summers of fluctuations, temperatures remained perfectly regulated even on scorching days. Employees and customers commented on the improved environment.

- **Saving Substantially on Bills:** Where electric bills had averaged $500 higher in summer months before, the first summer with the new unit came at $300 less, to the owner's delight. Projections showed a cost recovery within three years through energy savings. Lower bills provided valuable cash flow.

- **Word of Mouth Boost:** Pleased patrons asked about the improved environment, spurring restaurant referrals. Staff productivity also rose without debilitating indoor heat. Customer satisfaction increased along with new business from word-of-mouth advertising.

HOME EFFICIENCY UPGRADES: A FAMILY'S DRY CLIMATE SUCCESS

The Smith family resides in an arid Southwestern region where cooling dominates seasonal bills. After experiencing persistently high invoices, they took action.

- **Air Sealing and Insulation:** Contractors inspected for drafts and applied low-expanding foam plus fiberglass batts to seal attic bypasses and increase ceiling, wall, and floor insulation R-values. The project located and closed air leaks to prevent conditioned air from escaping.

- **Ductwork Repairs:** Cracked ducts in the non-conditioned crawlspace caused a 25% leakage rate. Mastic sealed all connections while new insulated ducts eliminated losses. Fixing duct leaks prevented wasted energy and improved airflow.

- **HVAC Tune-Up:** Service technicians replaced old capacitors and cleaned the evaporator coil and condenser coils outside, along with a full system checkup. A well-tuned system runs more smoothly and efficiently.

- **Thermostat Upgrade:** The outdated manual dial thermostat was overhauled with a smart WiFi programmable model featuring smartphone controls and greater scheduling flexibility. New controls enhanced comfort and energy management.

- **Window Replacement:** Single-pane windows gave way to more efficient dual-pane low-E glass versions, providing improved insulation. Reduced drafts and conductivity lowered cooling needs.
- **Landscaping Adjustments:** Xeriscape techniques, like native plant selection and drip irrigation, bounded yard sprinkler bills while beautifying the landscape naturally. Water-wise landscaping supplemented indoor efficiency efforts.
- **Energy Reduction Results:** After implementing the recommended upgrades, the Smiths savings hit 21% on electric bills the first year. With a dry climate dependent heavily on cooling before, their costs dropped substantially through comprehensive improvements.

RENOVATING A HISTORIC HOME'S HVAC SUITE

An antique farmhouse underwent modifications to modernize its antiquated systems while retaining period aesthetics.

Assessing Aging Equipment

Inspectors found the atmospheric gas boiler from 1938 badly corroded along with its uninsulated pipes. Rusted components posed safety and reliability issues. Additionally, rented wall heaters inefficiently supplemented insufficient heat. The antiquated arrangement was both unsafe and uneconomical to maintain. A full renovation was deemed necessary to bring the home into the present.

Plumbing Modern Upgrades

PVC pipes replaced aging metal ductwork throughout. The new piping offered better insulation compared to the old system. Zoned radiant floor heating was installed alongside five new radiators to distribute warmth evenly. Thermostatic valves provide precise temperature regulation from room to room. Water flow was balanced for maximum comfort.

Selecting an Efficient Heating System

A high-efficiency propane boiler and on-demand water heater were chosen for their clean-burning and energy-efficient operation. A hydronic system would offer quiet, consistent heat paired with instant domestic hot water. Programmable thermostats allow easy scheduling of heating and cooling cycles to maximize cost savings.

Preserving Historic Character

Exterior changes were minimized to maintain the home's vintage appearance. Modern components like the new furnace were concealed from view. Original wood trim, windows, and siding remained undisturbed. Interior ductwork was run through open floor joists, avoiding demolition. The renovation seamlessly blended functionality with historical accuracy.

Construction and Installation

Contractors first removed the aged gas boiler and ductwork before retrofitting piping in walls and floors. The new system installation required precision to avoid damaging antiques. The effort was made to shield surfaces from dust and debris. Within a few weeks, the project was completed with a fully functioning yet historically sympathetic HVAC suite.

Unparalleled Comfort Gains

Once activated, the upgraded system surpassed homeowners' highest expectations. Consistent, programmable warmth blanketed the entire abode for the first time. No longer did occupants battle hot or cold spots. The quiet operation allowed full enjoyment of the home without disturbance. Overall comfort and indoor environment dramatically benefited.

Delighted with Efficiency

Combined with insulation improvements, fuel bills decreased by 30% compared to previous winters. Performance far outstripped the unreliable and wasteful past arrangement. The cozy environment, long-term cost savings, and reliability accompany peace of mind. The home's antiquity blended seamlessly with modern convenience.

PART 5

ADVANCED HVAC INSIGHTS

PART 5

CHAPTER 9
Commercial HVAC Systems

Commercial HVAC systems differ greatly from residential systems due to the unique challenges of heating, ventilating, and cooling large commercial spaces. This chapter will examine the key differences between commercial and residential HVAC, discuss important considerations for sizing and designing commercial systems, explore strategies for improving energy efficiency in commercial settings, cover maintenance challenges and solutions, and provide real-world case studies of successful commercial HVAC implementations. Whether you own or manage a small business, retail operation, or large commercial building, the insights in this chapter will help you better understand and manage your HVAC needs.

Commercial HVAC systems are larger, more complex, and have higher capacity requirements than residential systems.

Size and Scale

Commercial HVAC systems are significantly larger than those used in homes due to commercial buildings' much greater conditioned space and occupancy. Air conditioning units for commercial spaces can range from 5 tons to over 100 tons of cooling capacity. Larger chillers, cooling towers, air handlers, ductwork, and pipes are also typically required.

The scale of commercial HVAC presents unique installation challenges. Lifting and maneuvering equipment into position requires heavy-duty cranes or lifting trucks. Ductwork also needs to be fabricated on-site due to its enormous size. Larger pipes may necessitate special tooling and welding skills during installation.

Maintenance and repairs also prove more difficult and costly than residential systems due to scale issues. Reaching and accessing high-mounted rooftop units or remote air handlers may require elevated work platforms for safety. Removing and replacing large composite cooling coils or centrifugal chillers entails substantial time, labor, and equipment usage. Breakdowns can be far more disruptive and expensive due to commercial HVAC equipment's scale.

Equipment costs also naturally escalate for commercial HVAC compared to residential units due to differences in size and capacity. A 15-ton rooftop air conditioner designed for a small office building may have an initial price tag several times higher than a residential 5-ton unit. Chillers, cooling towers, and other central plant equipment cost tens if not hundreds of thousands of dollars each. Large commercial HVAC projects can reach millions when all design, material, installation, and commissioning expenses are totaled.

Multiple HVAC Zones

Most large commercial buildings contain multiple HVAC zones to allow different building areas to be conditioned independently. For example, a multi-story office tower might have individual zoning for each floor. A large retail store could separate the sales floor from stockrooms. Manufacturing facilities commonly zone processing areas apart from administrative offices.

Careful consideration goes into defining thermal zones. Factors like occupancy levels, internal heat gains, ventilation needs, solar exposures, and more influence how a building should be partitioned. Dedicated HVAC equipment and controls are then required for each zone. A central plant often distributes heated/cooled water that terminal units or air handlers use within zones.

Zoning provides energy-saving benefits by allowing underutilized areas to be shut down after hours. It also enhances comfort control - the cafeteria space can remain comfortably cool while the warehouse warms naturally on weekends. The flexibility to switch between cooling and heating individual zones saves operating costs.

However, zoning adds complexity versus a single-zone approach. Distributed control systems must coordinate all zone equipment. Installation involves extra ductwork, piping, dampers, controls, and equipment to condition independent areas. Debugging and balancing multi-zone commercial HVAC can present challenges post-installation if not commissioned properly.

Higher Performance Requirements

Commercial HVAC systems experience greater demands than residential units due to having to condition space that may be densely occupied for long hours, have process loads from equipment, or need tight temperature and humidity control for specialized uses. This necessitates equipment with larger capacities, faster response times, tighter tolerances, and higher airflow requirements.

For example, an office building with dense cubicle layouts and large internal heat gains from people and equipment requires more capable air conditioning than a residential home of similar square footage. Spot cooling a restaurant kitchen during food preparation necessitates more responsive cooling than a household refrigerator. Humidity control for a hospital operating room demands higher precision than a typical home.

In such cases, conventional residential-grade HVAC equipment cannot deliver the level of performance needed for commercial applications. Larger chillers, cooling towers, air handlers, and other central plant equipment rated for commercial use are required. Premium components with closer setpoints, faster-acting valves, controls, and heavier-duty construction handle the load.

Achieving tight commercial indoor environment specifications also poses challenges. Factors like open plan offices, floor-to-ceiling windows, or process loads complicate conditioning large continuous spaces uniformly. Doing so requires careful system balancing, larger ducts/pipes for even airflow, redundant equipment for backup, and modern building automation.

Year-Round Usage

Unlike many homes, commercial facilities like offices, retailers, and manufacturing plants require conditioning 24 hours a day, 7 days a week throughout the year. There are no seasonal breaks for commercial HVAC equipment. Holidays and weekends mean continued operation. After-hours emergency calls necessitate spare parts and service staff availability at all times.

Non-stop usage places greater wear and tear on commercial HVAC systems versus periodic residential usage. Constant runtime puts more thermal and mechanical stress on components like chillers, cooling towers, air handlers, and large fans. Motors, seals, and moving parts degrade more quickly in this environment.

Often, lack of downtime for preventative maintenance also accelerates equipment deterioration. Waiting until a breakdown to service a unit results in further equipment damage and longer outages than periodically replacing filters, lubricating parts, and performing checks on an occupied schedule.

Year-round conditioning also drives up energy costs substantially. Commercial buildings naturally consume far more heating and cooling BTUs annually than residential homes, seeing part-time usage. Savings from optimizing systems, commissioning, and retrofits leverage larger potential returns due to constant commercial HVAC operation.

More Complex Control Systems

The advanced zoning, higher performance demands, and year-round operation of commercial HVAC necessitate sophisticated control systems. Programmable thermostats and controllers coordinate multiple air handlers, chillers, condensing units, and other equipment across all zones. Building automation systems directly modulate components and optimize runtimes based on various inputs.

Incorporate schedule programming, temperatures and setpoints, humidity control, demand control ventilation, alarm monitoring, energy management features, and compatibility with lighting, safety, and security systems. Proper programming and configuration of controls assume increased importance for commercial HVAC efficiency.

Technician skill levels must also rise to service complex automation. Troubleshooting distributed digital controls requires logical thinking skills, an understanding of schematics, and commissioning processes. Control sequences may involve PLC ladder logic, communicating protocols, and integrating sub-systems like BAS, BMS, fire safety, etc.

Importance of Air Quality

In commercial spaces where many people occupy indoor areas for extensive periods, air quality proves a much higher priority than most residential applications. Poor indoor air quality (IAQ) negatively impacts occupant health, comfort, and productivity. Consequently, commercial HVAC systems commonly address IAQ far more robustly.

Factors like microbial contaminants, off-gassing from materials, and indoor pollutants from processes/chemicals can escalate to serious concerns in offices and public buildings. Ensuring adequate outside air ventilation, filtration, demand controls, and indoor monitoring addresses such issues.

IAQ is also tied closely to building codes and regulations for commercial facilities. Standards specify requirements for ventilation, source control, humidity limits, and pollution monitoring or remediation. Meeting codes help control liability risks around indoor environments impacting occupants. Specialized spaces like hospitals necessitate even more stringent air quality management.

SIZING AND DESIGN CONSIDERATIONS

Proper sizing and designing commercial HVAC systems are crucial to condition spaces within performance requirements while minimizing energy use effectively.

Load Calculations

Accurate heating and cooling load calculations are the starting point for any commercial HVAC project. Variables like building envelope characteristics, occupancy levels, internal heat gains, ventilation requirements, and more must all be carefully accounted for.

Load calculation methodologies follow standardized procedures defined by organizations like ACCA and ASHRAE. Engineers perform manual calculations or use software programs to model the building, account for factors that impact loads, and size peak heating/cooling capacities.

Proper modeling incorporates building geometry, floor plans, wall/window/roof types, thermal properties, solar exposures, infiltration rates, and more. Internal gains come from occupants' body heat, lighting, plug loads from kitchens/offices, and process loads. Ventilation air volume depends on building codes and whether areas require 100% outside air.

Peak design conditions specify outside temperatures, whether the building operates continuously or seasonally, and indoor setpoint temperatures. Calculations determine peak heating/cooling capacities, process cooling needs, and ventilation air pre-conditioning loads. Detailed reports help select correctly sized equipment.

Load calculations require experienced analysts, as even minor assumptions/errors compound and can significantly impact equipment sizing. Undersizing risks insufficient capacity and discomfort complaints. Oversizing wastes costs and reduces part-load efficiency. Life-cycle costs directly relate to initial sizing, making load analysis critical.

Equipment Selection

Depending on building needs, utility costs, and budget, multiple equipment options exist for commercial HVAC. Choices include air-cooled or water-cooled chillers, rooftop units, split systems, geothermal heat pumps, boilers, and cooling towers.

Selecting the right equipment involves comparing first costs, efficiency ratings, capacities, dimensions/clearances needed, utility incentives, and compatibility with an optimized system design. Air-cooled chillers cost less initially but use more energy, while cooling towers enable more efficient water-cooled options.

Packaged rooftop units provide an easy installation but lack the individual component optimization of central plant designs. VRF heat pump systems offer zoning flexibility at a higher cost. Absorption chillers leverage waste heat as a free energy source.

Specialized process equipment serves unique loads from data centers, cleanrooms, industrial processes, etc. Configurations range from simple variable air volume arrangements to complex central chilled water plants with thermal energy storage.

Lifecycle cost analysis quantifies operating expenses to help identify the most economical combination of components meeting loads over the equipment lifespan. Modeling controls integration and part-load performance also impacts the best overall solution.

Air Distribution Design

The design of ductwork, diffusers, grilles, and air handlers is another important aspect. Airflow must be precisely balanced between zones to ensure uniform thermal comfort and avoid hot/cold spots.

Factors like duct sizing, insulation levels, leakage mitigation techniques, registers, filtration approaches, and whether VAV, constant volume, or other air delivery will be used require consideration. Software programs model airflow balancing through duct runs to minimize fan power.

Using properly sized ducts shortens lengths and reduces friction losses, saving fan energy costs significantly over the long run. Mechanically fastened and sealed duct joints prevent leakage of costly conditioned air. Filters remove particulates for indoor air quality while minimizing resistance.

Placing diffusers/registers directs airflow patterns to sweep thermal zones effectively while preventing drafts. Factors like plenum designs, hanging considerations, noise criteria, and surface finishes impact comfort. Proper air distribution boosts system performance and occupant satisfaction.

Controls Layout

Building automation systems, large unit controllers, and thermostats necessitate a carefully planned control strategy to maximize efficiency and comfort. Key design components include sensor placement, setpoint scheduling, interfacing between equipment, optimizing operation sequences, and energy management capabilities.

Factors like HVAC enable/disable scheduling, chilled water pumping schedules, VAV/CRAC unit interaction, ventilation requirements, and building pressure controls all impact sequences. Trend logging and remote access features allow performance monitoring and fault detection.

Controller Placement considers wiring runs and conduit routing planned from field devices back to central control panels. Communicating protocols dictate compatible controllers and gateways for interfacing legacy equipment. Contingency plans address equipment failure scenarios.

Controls directly impact occupant experiences and operational costs through optimized sequences and the ability to remotely commission, maintain, and retrofit systems for continuous improvement. Detailed planning sets projects up for control success.

System Placement

Rooftop units, cooling towers, air intakes/exhausts, and other exterior equipment need ideal locations for functionality, service access, and aesthetic/noise factors. Indoor air handling units, chillers, and system components require considering space constraints, structural considerations, and installation logistics during layout.

Factors like structural loading capacities, wind loads, seismic bracing, fall protection, and rooftop weight distribution/routing of utilities to interior mechanical spaces influence optimal component placement. Indoor clearances allow servicing, while outdoor enclosures address acoustics/visual screening.

Convenience also affects layout. Central plant designs locate chillers/boilers/towers together for easier maintenance versus distributed RTUs. Spare parts storage considers replacement component movements. Future expansion capabilities guide initial designs.

Commissioning and Testing

Upon installation, commercial HVAC systems require thorough commissioning, testing, and balancing to ensure design parameters are met before client acceptance.

Air/water flow, temperatures, pressures, component functionality, control sequences, and overall integrated performance across zones must all be rigorously evaluated. Training occupants certifies proper system use. Any issues identified can then be corrected for trouble-free operation.

Commissioning validates installation quality contractor artistry and calibrates automated control sequences through seasonally comprehensive testing. Deficiencies remedied under warranty help ensure long-term reliability. Retainage withholds final payment pending full commissioning completion.

ENERGY EFFICIENCY IN COMMERCIAL SETTINGS

As HVAC comprises a large portion of commercial building energy usage, optimizing operation for efficiency can substantially reduce costs while improving environmental impact. This section examines strategies for boosting the performance and sustainability of commercial HVAC systems.

High Efficiency Equipment

ENERGY STAR or similarly rated equipment offers significant, cost-effective efficiency gains versus standard models. For new construction, selecting chillers, boilers, fans, and other components with the highest SEER, EER, AFUE, or similar energy factor ratings pays long-term dividends through utility cost savings.

Higher efficiency equipment costs marginally more upfront but provides quicker returns through reduced consumption. Lifecycle analysis finds that savings far outweigh additional initial expenses over the equipment's lifetime. Replacing outdated, lower-efficiency units makes sense if the replacement payback time is three years or less.

Manufacturers continuously advance commercial HVAC technology, so keeping purchase specifications up-to-date leverages the latest offerings. Incentive programs sometimes subsidize efficient choices, further strengthening their value proposition. Referencing databases like AHRI or the ENERGY STAR qualified product list simplifies comparing performance ratings.

Active Controls

Optimizing HVAC through advanced control sequences and strategies captures substantial savings compared to uncontrolled operation. For example, adjusting chilled water temperatures up a few degrees when the building is lightly occupied reduces chiller energy significantly.

Automation allows tighter temperature control, minimizing simultaneous heating and cooling from control headbands. Modulating control valves and variable speed drives optimize fan and pump speeds based on real-time demand rather than constant maximum flow.

Pre-cooling or pre-heating outdoor air with economization or enthalpy wheels helps downsize mechanical equipment. Demand control ventilation monitors carbon dioxide and adjusts minimum ventilation rates to actual occupancy levels.

Retrofitting older controls provides an excellent upgrade path. Updating pneumatic systems to DDC, adding occupancy sensors, integrating lighting controls, and remote monitoring access shapes controls into an energy management system. Commissioning fine-tune sequences.

Heat Recovery Systems

Capturing normally wasted thermal energy presents large commercial savings potential. Applications include heat recovery chillers, desiccant dehumidification, runaround coil loops, thermal wheels, and energy recovery ventilators.

Heat recovery chillers use a second circuit to pre-heat domestic water or pre-condition outside air. Runaround coils transfer waste heat between distinct airstreams like outdoors and a data center exhaust. Rotating thermal wheels simultaneously pre-cool incoming and reheat outgoing airstreams during ventilation.

Such equipment exploits overlapping heating/cooling needs common to commercial facilities. Integrating with existing HVAC yields paybacks under three years typically. Modeling system optimizations uncover the most promising waste heat recovery applications.

Economizers

Economizer cycles harness "free cooling" potential via enthalpy controls air or water side economizing. On milder days, bypassing mechanical refrigeration lowers electrical demand charges and consumption costs substantially compared to compressor operation alone.

The percentage of annual hours an economizer can operate varies regionally but often exceeds 2,000 hours annually. Modeling climate data determines optimal cutover control sequences. Modern digital controls maximize economizer usage safely.

Equipment mounting positions consider wind-driven rain intrusion risks. Backdraft/relief dampers address building pressure concerns. Filter racks protect coils from debris drawn in during free cooling. Proper maintenance ensures economizers reliably activate when conditions permit.

Insulation and Air Sealing

Applying sufficient insulation to piping and ductwork prevents thermal losses from distributing conditioned air and water throughout buildings. Ensuring tight duct sealing safeguards even more energy by stopping uncontrolled air leakage.

Codes specify minimum R-values for piping based on diameter and fluid temperature. Meeting or exceeding norms yields compounding savings over many years of use. Mechanically fastened and taped seams create durable, long-lasting, sealed duct systems.

Locating and addressing major air leaks through pressure diagnostic testing achieves the greatest savings potential from sealing efforts. Re-insulating older, inadequate systems provides a quick payback. Continuous commissioning monitors insulation/sealing quality over the long run as well.

Lighting/Daylight Harvesting

Commercial building efficiency extends beyond HVAC through integrated design. Proper lighting design leverages high-efficacy LED luminaires, occupancy sensors, daylight harvesting, and customizable controls.

Optimized lighting placement and dimming strategies minimize energy usage while delivering task illuminances. Daylight sensors automatically adjust artificial lighting based on natural light availability. Customizable controls enable independent scene settings.

Integrating with HVAC occupancy scheduling and temperature setbacks yields compound savings. Coordinating lighting/HVAC projects captures interactive energy reductions unachievable through isolated initiatives. Installing efficient lighting also lowers internal heat gains and HVAC equipment sizing requirements.

Renewable Energy Integration

Renewable sources like solar thermal panels, photovoltaics, biomass boilers, geothermal, and wind can offset peak HVAC loads or provide a portion of base building needs through green power generation.

Solar hot water helps meet domestic or space heating demands. PV panels generate electricity to directly power HVAC equipment or export excess to the grid. Thermal energy storage decouples generation from end uses.

Modeling renewable feasibility considers incentives, roof access, zoning codes, and long-term energy cost avoidance. Right-sizing systems optimize renewable capacity while avoiding oversizing. Integrating diverse renewables expands benefits.

Prioritizing high-efficiency equipment, optimized controls, waste heat recovery, economization, sealing/insulation, coordinating HVAC/lighting, and leveraging renewables establishes a pathway for substantial commercial energy reductions and cost savings. Comprehensive efficiency strategies lower consumption through operation optimization.

MAINTENANCE CHALLENGES AND SOLUTIONS

Keeping commercial HVAC systems operating reliably and efficiently poses unique challenges versus residential maintenance. This section discusses common issues that arise and proven solutions for continuous optimal performance.

Continuous Operation

24/7/365 runtime places greater mechanical and thermal stress on commercial HVAC equipment. Components like motors, seals and bearings deteriorate more rapidly without seasonal breaks. Adding to wear are constant start/stop cycles controlling multiple zones.

To combat accelerated equipment wear, thorough preventative maintenance becomes paramount. Whereas a home system may receive seasonal checkups, commercial PM involves more frequent filter changes, belt inspections, oil changes, and comprehensive seasonal services.

Keeping maintenance records centralizes component service histories and facilitates predicting and planning for replacements before critical failures. Budgeting for PM programs safeguards uptime while extending asset lifecycles cost-effectively.

Complex Component Access

Reaching all areas of large, height-mounted commercial HVAC units often requires boom lifts or scaffolding for inspection and service. At the rooftop, falling hazards demand harnesses and adequate work platforms per safety regulations.

Similarly, accessing indoor air handlers or basement-level chillers/boilers may involve navigating tight or restricted mechanical spaces. Special lifting gear removes heavy parts like cooling coils or fans securely.

Technicians require specialized tools for component removal/replacement. Tasks like changing retainer clips or separating piping connections on major equipment demand proper implementation. Inventorying such specialty tools streamlines difficult access maintenance tasks.

Specialized Skills

Troubleshooting complex sequences on distributed control systems and communicating building automation demands control programming skills. Servicing premium components and specialty process cooling units involves capabilities beyond general technicians.

Cross-trained staff properly licensed, educated, and experienced on diverse commercial equipment optimizes problem-solving and uptime. Outsourcing specialized maintenance until in-house expertise develops risks, delays, and equipment damage from improper servicing.

Critical Nature of Operations

HVAC outages disrupting commercial business operations carry higher costs than residential comfort losses. Backups assure continuous conditioning for mission-critical facilities.

Redundant equipment, automatic control transfer, and fast-response service contracts mitigate downtime risks. Parts inventories shorten MTTR for priority-one equipment. Planned seasonal maintenance minimizes unexpected breakdowns interrupting occupancy.

Tight Indoor Environments

Commercial indoor conditions like healthcare aseptic areas, cleanrooms, or data center environments require ultra-precise temperature/humidity control and stringent air quality. Even minor excursions threaten processes or health.

Redundant precision HVAC components ensure tight tolerances. Robust control and monitoring systems detect drifts immediately for correction. Frequent coil/filter changes and sterilization prevent microbial threats within sensitive spaces. Maintaining certification takes experienced technicians.

Inter dependencies

Integrated building systems interlinking HVAC with other disciplines like lighting, security, or BMS complicate troubleshooting. Coordinating maintenance windows minimizes cross-system outages.

Documenting all integration points streamlines change management and system updates. Keeping automated sequences in sync requires control skills across specialties. Periodic testing validates interface functionality remains intact over time.

Changing Codes & Standards

Evolving regulations and guidelines necessitate periodic system and process updates. Technicians pursuing continuing education keep awareness of new requirements for air quality, efficiency, indoor environments, or materials like refrigerants.

Capital planning budgets for code-driven retro-commissioning, balancing, re-verification, or modifications to optimize evolving standards long-term. Proactive adoption maintains compliance safely without reactive hurdles.

Preventative Maintenance

Routine inspections and servicing form the backbone of any commercial maintenance program. Key elements include filter changes, belt adjustments, oil services, cleaning, and coils/fans/motors/pumps/valves inspection.

PM program implementation crucially involves documentation, scheduling of routine tasks, parts/supplies stocking, performance of checks by trained technicians, and recordkeeping of maintenance history in CMMS. Proactive execution avoids disturbing breakdowns.

Training & Safety

Commercial HVAC work requires adherence to lockout/tagout procedures due to high voltage/pressure/temperature hazards. Technicians hold proper licensing/certifications and complete ongoing safety training.

Developing in-house expertise occurs through cross-training, mentoring newer hires, reimbursing continuing education, and tracking competency development. Documentation demonstrates due diligence if injuries occur. Prioritizing safety supports compliant operations.

Addressing issues like continuous run time, access challenges, specialized skills needs, critical reliability, tightly controlled environments, integrated systems, evolving codes, and comprehensive preventative maintenance programs establish excellence in commercial HVAC and building operations support. Proactive strategies maintain peak performance cost-effectively.

Case Studies in Commercial HVAC Success

Real-world examples demonstrate proven strategies for implementing successful and efficient commercial HVAC systems. This final section provides two case studies highlighting best practices.

Office Renovation Boosts IAQ and Cuts Costs

A 30,000-square-foot office building underwent renovations to replace its dated HVAC system, struggling with indoor air quality and high utility bills. Engineers designed zoning, installed VARIABLE REFRIGERANT FLOW heat pumps, implemented building automation for demand-controlled ventilation based on CO_2 levels, and fitted the space with occupancy sensors and LED lighting. Results included a 15% reduction in energy usage, improved employee productivity, and no sick days attributed to indoor conditions since the retrofit.

Green Retrofit for Retail Chain

A national retail chain invested in energy-saving HVAC upgrades across 300 stores averaging 50,000 square feet each. Chillers were replaced with water-cooled variable speed units featuring improved part-load efficiency. New VAV air handlers included economizers and EC fan motors. Controls integration allowed immediate remote monitoring and adjustment of setpoints nationwide. Combined projects paid for themselves in under two years via reduced utility costs, resulting in yearly savings exceeding $3 million while slashing the company's carbon footprint.

Delivering Comfort via Geothermal

A LEED Gold-certified office complex tapped geothermal technology for environment-friendly HVAC. A closed-loop well field and water-to-water heat pumps provide year-round space conditioning and domestic hot water. Careful design and construction involved instituting water treatment, air distribution modeling, rigorous testing to ensure 40 gpm groundwater flow, and modern building automation. Geothermal HVAC helped the facility achieve net-zero status through ultra-high efficiency and renewable energy generation onsite.

Retro-commissioning Pays Off

Faced with fluctuating temperatures and humidity complaints despite equipment upgrades, a manufacturing plant performed retro-commissioning of its HVAC systems. Tuning and calibration of controls identified issues like improperly configured static pressures causing terminal unit air starvation. Resetting setpoints and equipment schedules optimized fan operation. Repairing steam leaks and tightening damper linkages resolved comfort problems. Annual savings of $75,000 materialized through 18% energy reduction and improved productivity.

University Campus Goes Green

A large university faced piecemeal campus growth, increasing HVAC energy usage. Its master plan incorporated consolidating smaller systems into three central energy plants supplying thermally zoned buildings via an underground piping distribution network. Variable primary flow optimizes pumping energy. Heat recovery chillers provide hot/chilled water using 30% less primary energy. Integrating renewable sources like a 2.5-megawatt solar array and fuel cells will help the campus achieve carbon neutrality.

These real-world case studies illustrate definitive methods for tackling commercial HVAC challenges through optimized design, commissioning, retrofits, central plants, alternative technologies, integrated control systems, and renewable energy. Demonstrating proven solutions inspires replicating success in other commercial facilities.

CHAPTER 10

HVAC Regulations and Environmental Impact

The refrigeration and air conditioning industry operates within an extensive regulatory framework to protect public health, safety, and the environment. Understanding, navigating, and staying compliant with these regulations is crucial for HVAC professionals and businesses. This chapter provides an overview of the current regulatory landscape and how HVAC systems impact the environment. It also explores green technologies being adopted to reduce emissions and improve sustainability. With greater knowledge of regulations and a focus on environmentally friendly practices, readers can help mitigate costs through increased efficiencies and avoid compliance issues.

Regulatory Bodies

The main regulatory bodies that establish HVAC standards in North America are crucial to understand for compliance. The Environmental Protection Agency, or EPA, regulates environmental issues, like refrigerants and other pollutants emissions, through acts like the Clean Air Act. On a state or provincial level, some agencies oversee licensing requirements for HVAC contractors and technicians. Things like certification, permitting, and code adherence fall under their purview. State agencies also promote public safety regarding HVAC installation and maintenance standards.

Some key state agencies include the New York Department of State's Division of Building Standards and Codes, the Texas Department of Licensing and Regulation, and California's Contractors State License Board. On a more local level, cities and counties typically enforce construction codes set forth by organizations like ICC and NFPA. This involves things like permitting for new builds and major renovation projects. Utilities also play a role through energy efficiency programs and incentives. Overall, HVAC professionals need to understand the scope and responsibilities of the various regulatory bodies to avoid issues of noncompliance down the line.

Installation Standards

Designing, installing, and modifying HVAC systems must adhere to standards set forth by ACCA (Air Conditioning Contractors of America) and ASHRAE (American Society of Heating, Refrigerating and Air-Conditioning Engineers). These cover best practices for equipment sizing, duct design, load calculations, safety features, insulation requirements, and ventilation specifications. Proper documentation of a system's design is crucial should any issues arise after project completion.

Compliance with standards set by groups such as the International Residential Code (IRC) and International Mechanical Code (IMC) is expected for residential installation projects. This includes regulations about proper equipment support, condensate drainage/disposal methods, combustion air supply, exhaust venting, and termination clearances. Detailed load calculations must justify equipment selection and confirm adequate capacity for meeting a home's heating/cooling needs under local climate conditions.

On the commercial side, standards published by ASHRAE and the International Building Code (IBC) are typically referenced. These deal with load determination approaches, system component specifications, indoor air quality necessities, energy conservation mandates, and documentation submittal procedures. Proper licensing is also needed as commercial HVAC systems involve greater capacities, pressures, voltages, and refrigerants, presenting higher safety risks if not installed correctly.

Permitting Requirements

Securing the necessary construction permits from local authorities is an important compliance step, allowing building inspections to check for code conformity. Residential permits are usually required for the installation of new HVAC equipment as well as major replacement/renovation jobs that entail ductwork modifications or structural alterations. Typical commercial projects that demand permitting and inspections include new buildings, tenant build-outs, equipment swaps, and remodeling/expansion undertakings increasing a structure's conditioned area.

Permit applications must include specifications for the contemplated HVAC materials, a system design diagram, and evidence that installation will adhere to existing codes. Some jurisdictions additionally require documentation of appropriate licensing for procuring mechanical consents. The permit process exists to confirm safety and installed performance as defined by standards - issues that can become compliance matters if overlooked. Contractors are wise to allow sufficient time for obtaining permits and scheduling all obligatory rough-ins and final inspections.

Record-Keeping Requirements

Robust record-keeping serves several critical compliance functions. It allows HVAC businesses to demonstrate conformance to regulations if asked by an authority. Documentation of permitted installation projects, equipment service/maintenance activities, employee training, and refrigerant management transactions provides a historical record should any problems or liability issues emerge.

Service records must capture equipment specifications, control settings, inspection findings, replacement part numbers, refrigerant pressures/temps, client signatures, and tech identifiers. Refrigerant transaction certificates proving lawful acquisition, usage, and disposal are also important compliance paperwork. Documentation also assists in verifying work was done correctly to industry standards and manufacturer specifications.

Organized physical or digital filing systems help access necessary records promptly for compliance proof, especially if an official inquiry occurs years later. Keeping tax documents in order can further benefit during audits by justifying standard operating expenses and tax deductions. Compliance-minded documentation offers protection and often saves time, clearing up misunderstandings compared to incomplete or missing records.

Ongoing Training Requirements

Many HVAC licensing certificates imposed by state authorities mandate accruing continuing education credits on an annual or biannual cadence. These CEU requirements uphold technician competency in light of always-evolving technologies, best practices, safety procedures, building codes, and environmental regulations. Compliance is verified during license renewal by presenting proof of completed training courses.

Online or in-person classes from providers like Refrigeration Service Engineers Society (RSES), Air Conditioning Contractors of America (ACCA), and North American Technician Excellence (NATE) offer training pertinent to regulatory updates. Course topics may encompass refrigerant management, ventilation systems, electrical safety, efficiency standards, pollution mitigation tactics, and codes/standard revisions.

Additionally, manufacturers periodically issue required training on new products, upgrades, installation/service techniques, and legal/warranty implications. Leveraging employer-provided or subsidized courses satisfies licensing needs while gaining skills promoting career success. Non-compliance with continuing education mandates may incur penalties or prevent relicensing if audit nonconformities surface.

Energy Efficiency Regulations

Given HVAC's substantial electricity consumption worldwide, governing bodies impose minimum efficiency benchmarks to curb energy waste and associated emissions impacts. Residential appliances like air conditioners, heat pumps, furnaces, and boilers must meet the Department of Energy's (DOE) efficiency standards, which are strengthened regularly per the Energy Policy and Conservation Act. Regional standards may impose more stringent specifications, too.

For large commercial/industrial equipment classes spanning chillers, unit ventilators, and package terminal air conditioners, renewed efficiency mandates subject to ASHRAE/IES (Illuminating Engineering Society) Standard 90.1 and later progressed to ASHRAE Standard 90.1 take precedent nationally. Additional specifications exist around economizers, duct sealing, static pressures, and control capabilities.

Compliance involves checking AHRI certification ratings to verify that a product model surpasses prescribed efficiency thresholds before sale or installation. Field inspections can cross-reference equipment nameplates with AHRI directories as well. Theft of regulated HVAC components poses regulatory noncompliance and associated penalties. Permits necessitate efficiency confirmation, too.

Energy consumption regulation extends to building envelopes. Insulation minimums, window standards, lighting power allowances, and control requirements aim to reduce HVAC-influenced loads stipulated within IECC and ASHRAE 90.1 for commercial buildings. Overall, energy modeling proves conformance.

Meeting efficiency regulations yields long-term cost savings while slashing emissions economy-wide. Compliance requires keeping abreast of strengthening metrics and incorporating the highest-rated equipment into designs whenever modifications occur. Strict adherence cuts liability risks and carbon footprints.

Special HVAC Situations Requiring Regulatory Consideration

Certain industries, like data centers, present unique regulatory needs, given extreme sensitivity to temperature/humidity control variances. Federal acts mandate stringent upkeep of temperature/humidity logs, demonstrating special cooling equipment reliably maintains service levels despite abnormal weather events.

Emergency power systems essential for healthcare, government, and public safety facilities necessitate life-safety grade HVAC functionality. Compliance entails emergency generator capacity verifying airside loads support crucial ventilation throughout blackouts. Transfer switch approvals affirm seamless, code-sanctioned HVAC transition onto backup power.

Marine vessel HVAC arrangements require Coast Guard preapproval to validate safe and lawful operations. Refrigeration systems aboard ships presenting foodborne illness risks necessitate precautionary design confirming reliability under hazardous conditions per the FDA food code.

Indoor agricultural cultivation undertakings may fall under the Department of Agriculture's oversight of managed-environment farming practices. HVAC permitting substantiates system capabilities to support crop production safely without the risk of illegal substances. Local jurisdiction approvals also factor in for these specialized operations.

Mission-critical computer rooms rely on HVAC uptime assessed under stricter tier classifications defined by the Uptime Institute. Compliance combines reliable redundancy, intense monitoring, fast automated controls, strict change management, plus availability stipulated by data center service level agreements.

Non-standard HVAC applications tied to special industries elicit unique compliance considerations beyond typical residential/commercial codes. Interdisciplinary coordination ensures lawful, low-risk solutions validating regulatory specifications from all applicable agencies.

ENVIRONMENTAL IMPACT OF HVAC SYSTEMS

Refrigerant Management

Refrigerants pose an ozone depletion and global warming threat if released into the atmosphere. As such, the EPA comprehensively regulates their usage, storage, transportation, and disposal through the Significant New Alternatives Policy (SNAP) program. This review process evaluates emerging low-impact alternatives to banned or heavily restricted refrigerants.

For example, HCFC phase-outs mandated the R-22 refrigerant replacement with environmentally preferable blends like R-410A. Further HFC phasedowns now necessitate next-generation alternatives such as natural refrigerants or lower-GWP synthetic options providing comparable efficiency. Proper EPA-approved safety training and certification holds particularly for flammable alternatives.

Compliant refrigerant procurement requires maintaining transaction logs verifying amounts, sources, and end-use situations. Refrigerant recovery for reuse or destruction during equipment decommissioning forestalls unlawful venting harmful to sustainability goals. Refrigerant sales or transfers mandate establishing purchaser identity and intended applications.

Leak detection and repair under §615 of the Clean Air Act cuts emissions at operational facilities. Many jurisdictions impose maximum leak rates for stationary refrigeration systems. Furthermore, the adequate quantity for system charging and service needs develops into a compliance consideration to prevent overfilling or supply shortages impacting repair schedules and costs.

Emission Regulations

Air pollutants detrimental to environmental and human wellness fall under emission standards purview. The EPA restricts particulate matters like dust under 40 CFR Part 50 for stationary cooling towers and refrigeration devices. Direct equipment emissions face changing requirements as effectiveness improves over time.

Manufacturers must certify that HVAC/R and refrigeration equipment attains current emission level thresholds tied to its capacity tonnage using standardized ARI/ASHRAE emission measurement protocols. Compliant loose parts prohibitions forestall rather than repair excessive ozone-depleting substance venting.

Additionally, §612 of the Clean Air Act empowers the EPA to set service practice standards preventing intentional refrigerant releases during repair, servicing, or disposal of refrigeration/AC machinery containing Class I or Class II substances. Compliance aids in meeting climate targets.

Energy Efficiency Standards

Given HVAC's energy intensity, the Department of Energy promotes cost-effective efficiency boosts minimizing environmental impact. Residential appliances respect standards issued in final determinations every six years as required under 42 U.S.C. 6295, while commercial equipment complies with ASHRAE 90.1 specifications.

Two thousand nineteen final rules bolstered standards for furnaces, air conditioners, heat pumps, water heaters, and other residential appliances. Commercial prerequisites ratcheted up for packaged and split-system air conditioners, heat pumps, computer room air conditioners, variable refrigerant flow systems, unitary and applied heat pumps, warm-air furnaces, commercial package boilers, and more.

Certification through the AHRI Directory proves HVAC systems ship at or above the minimum energy performance levels set forth by current DOE test procedures and efficiency criteria. Field verification cross-references nameplate data. Noncompliant, non-certified equipment faces prohibition, whereas substitutes deliver outsized benefits, cutting consumer costs.

Building Energy Codes

International, state, and local building codes target reducing total energy consumption and emissions from residential and commercial buildings using prescriptive and performance-based compliance paths. These regulate building envelope metrics, including insulation values, window characteristics, air leakage standards, and rooftop prerequisites.

A shortage of envelope compliance poses HVAC sizing, efficiency, and operational cost consequences. Beyond reinforced insulation mandates, codes increasingly necessitate on-site renewable integration and commissioning processes to verify performance matches specified design targets. Third-party inspections validate conformity with stretch code options available.

ASHRAE 90.1-2019 strengthened commercial envelope provisions, emphasizing wise water usage, responsible refrigerant selection, and smart control systems. Compliance necessitates confirming total building energy performance through modeling or equivalent methods. Rigorous acceptance testing verifies performance as designed across various climate zones.

Environmental Permitting

Larger construction projects may fall under environmental assessment requirements through state and local administrative authorities exercising National Environmental Policy Acts (NEPA). These evaluate impacts on air, water, land, wildlife, vegetation, noise, plus cultural/historical resources.

HVAC system design constitutes part of constructability reviews, ensuring solutions avoid or mitigate environmental harm. Carbon footprint modeling justifies system selections. Offsets balance any residual effects to attain development permission. Decommissioning and recycling plans also factor into permitting responsibilities.

Overall compliance embraces renewable-forward choices, energy-water nexus considerations, climate-adaptive strategies, and ongoing commissioning, maintaining design efficiencies and reducing environmental impact over decades of operation. Agencies leverage such foresight, strengthening sustainability locally.

GREEN HVAC TECHNOLOGIES: SUSTAINABILITY FOCUS

High Efficiency Equipment

HVAC manufacturers rapidly improve equipment to meet tighter efficiency benchmarks cost-effectively. Variable speed compressors deliver capacity modulation for matched airflow without parasitically consuming power at part loads. Inverters facilitate the same outcome within heat pumps.

Advanced heat exchangers boost coil performance using microchannel technology, Rifled Tube designs, or MicroGrooved tubing. Electronically Commutated Motors (ECM) provide variable speed control to air movers while cutting one-third of the energy usage relative to shaded-pole motors. Ductless mini-splits utilize multiport tube expansions, minimizing refrigerant quantities by twenty percent.

Buildings outfitted with the latest high-efficiency equipment achieve major savings exceeding energy code minimums. Heat pump water heaters cut utility bills versus resistance water heating as well. Tiered incentives and falling costs accelerate widespread market adoption, lowering emissions. System optimization using economizers, Demand Control Ventilation (DCV), and optimum start/stop enhance efficiencies.

Even retrofitting aging equipment with efficiency upgrades like ECM blowers or variable speed heading boosts yields quantified benefits. Integrating smart home features permits demand response potential through remote system control. Expanding building commissioning identifies further optimization tweaks, keeping systems operating at peak performance over the long haul.

Renewable Energy Integration

Renewable generators meet building loads cleanly by displacing grid power reliant on carbon fuels. Rooftop solar photovoltaic panels transform sunlight directly into electricity, offsetting HVAC consumption or feeding excess back for revenue. Fuel cells run emission-free on natural gas.

Once only feasible for utilities, micro-wind turbines now support small commercial properties, while geo-exchange systems leverage constant earth temperatures for year-round HVAC. Thermal energy storage improves renewable economics by buffering intermittent sources.

HVAC equipment and on-site batteries store renewable generation for later use, curtailing demand charges. Integrated control systems optimize all distributed energy resources for demand response participation, reducing peak loads and saving additional money through demand management rebates and time-of-use rates.

Manufacturers roll out product lines, simplifying renewable leveraging at any size. Plug-and-play modular solar packages turn roof spaces productive. Integrating renewable credits and incentives into HVAC project budgeting strengthens investment returns further.

Demand Response Enrollment

Linking building energy management systems and HVAC equipment to utility demand response programs cuts costs through participation incentives. Periodic load shedding avoids high energy pricing during grid peaks driven by temperature extremes or generation constraints.

Demand limiting during curtailment periods may involve duty cycling equipment like air handlers or chilling loads for a short duration to trim Usage but maintain occupant comfort. Facility lighting and plug loads assist as well under automatic control. Thermostats permit set point adjustments that are barely detectable.

Commercial and industrial facilities benefit from reducing summer demand charges by thirty percent using automated curtailment strategies. Duty cycling HVAC ten minutes on and twenty minutes off reduces equipment wear, extending lifespan, too. Virtual power plants aggregate distributed energy flexibility for grid services.

New products emerge, simplifying demand management through cloud-based virtual endpoints installed within minutes. Turnkey solutions optimize building systems, tapping considerable demand flexibility as a grid resource compensated as a utility asset. Demand response potential grows with technology access by all property types.

Zero Emission HVAC Options

Electrification promotes substituting combustion HVAC in favor of renewable-integrated high-efficiency heat pumps delivering heating, cooling, and often domestic hot water from a single unit. These use reversible cycle technology, concentrating refrigerant heat exchange for temperature adjustment rather than relying on carbon emissions.

Ductless mini-split heat pumps serve small, retrofit spaces adeptly. Larger packaged terminal and variable refrigerant flow heat pumps meet diverse commercial loads, too. Absorption chillers powered by waste heat or natural gas provide alternative carbon-neutral cooling.

Geothermal heat pumps exchange thermal energy between underground groundloops and buildings for whole building coverage. These distribute climate control through heated/chilled water and exhibit the HVAC industry's highest efficiency at any scale.

Renewable-powered heat pumps maximize savings through emissions avoidance and building electrification. Pairing these with on-site solar makes zero-emissions HVAC ubiquitous, supplanting direr climate impacts from fossil dependence over the long term.

Advanced System Optimization

Commissioning newly installed or retrofitted HVAC systems ensures performance matches design through operational verification and equipment calibration. Re-commissioning identifies optimization tweaks maximizing efficiency as buildings change over decades.

Continuous commissioning uses sensor analytics detecting variances from baseline baselines, showing savings opportunities like additional trims to ventilation or schedule adjustments. Retro-commissioning recommendations may include control system upgrades, economizer repairs, or duct sealing.

Equipment and building tune-ups regain lost efficiency from degradation and reduce energy waste from thermal bridging or air leakage. Optimized schedules conserve further occupancy. Monitoring verifies persistence while highlighting needed maintenance like filter replacement or coil cleaning.

Ventilation heat recovery systems capture outgoing air's thermal energy for incoming airstreams using heat pipes, heat wheels, or run-around coils. Extracted value supports 100% outside air, even in extreme climates. Thermal energy storage shifts loads by pre-cooling/heating buildings with chilled/hot water while avoiding peak periods.

FUTURE REGULATIONS AND INDUSTRY TRENDS

Refrigerant Phase-Downs

Heightened restrictions target thinning production and consumption of remaining ozone-depleting substances by 2030 under the Kigali Amendment. Upcoming HFC phasedowns also motivate embracing climate-friendly alternatives given powerful global warming potentials.

Transitioning the vast installed base presents logistical challenges outweighed by environmental benefits over the long haul. Technicians require extensive certification for new blends handling differently than legacy chemicals. Detailed recordkeeping substantiates compliant phase-in/refrigerant swaps spanning decades.

Manufacturers reformulate equipment utilizing low-GWP refrigerants, safely sustaining performance. Sufficient industry buy-in accelerates widespread applicability and cost reductions through innovation. Broader refrigerant availability and competitive options further ease regulatory transitions maintained under subsequent laws.

Net-Zero Carbon Buildings

Growing stakeholder commitments envision carbon neutrality by 2050, necessitating decarbonization strategies across all sectors. Building codes assist through energy waste prohibitions and renewable requirements. HVAC design focuses on eliminating fossil fuels by leveraging highly efficient heat pumps, geothermal exchange, and fuel flexibility capable of drop-in biofuel utilization.

Integrated renewables, storage, and demand flexibility optimize emissions profiles. Commissioning, monitoring, and re-tuning retain peak efficiency as a living building adapts to usages over time. Policy nudges carbon pricing internalization, changing project economics favoring the lowest carbon options. Low-emissions contractor credentialing recognizes sustainability leadership within supply chains.

Climate Change Mitigation Policies

Regulated carbon markets establish prices guiding long-term investments. Mandated emissions reporting raises accountability as climate impacts intensify, calling for bolder action. Carbon reduction planning quantifies mitigation opportunities across sectors informing policy development. Performance standards ratchet tightened, driving innovation and filling the technical feasibility gap.

Building codes incorporate climate resilience, necessitating durable HVAC functionality across severe events. Emergency planning integrates distributed energy resources supporting critical services. Land use adapts urban heat islands and expands natural carbon sequestration through greener cities. Electrified buildings leverage clean, dispatchable generation rather than emitting assets risky under climate uncertainties.

Decarbonization of Electricity Grids

Rapid renewable energy scale-up transforms centralized fossil fuel generation toward distributed clean resources supported by smart infrastructure. Time-varying carbon intensities factor into load shaping, enabling decarbonized HVAC operations. Interconnectivity permits the dispatching of flexible building loads, assisting variable wind and solar integration onto enhanced transmission networks.

Two-way power flow and integrated storage optimize distributed energy resource coordination. Transition assistance programs smooth disruption from declining emitting assets and carbon-intensive industries. Open standards encourage interoperability of customer-sited technologies participating as grid assets. Performance-based incentives motivate emissions reductions from all grid-edge technologies, including high-efficiency HVAC.

Increasing Resiliency to Climate Disruption

Extreme weather damages critical infrastructure, requiring climate-adaptive reinforcement and backup power capabilities. HVAC assumes expanded roles prioritizing occupant safety, public health, and community services functionality throughout disruptions. Reliable operation necessitates mitigation against compound events exceeding historical norms.

Durability upgrades withstand intensifying tropical storms and wildfires. Microgrids support temperature-controlled facilities isolation from wider grid disturbances. Smart controls configure HVAC for energy-conservative modes during prolonged outages to preserve limited on-site generation. Renewable-hybrid systems provide indefinite resiliency, whereas standardized interconnections streamline coordinated community response.

Proactive planning quantifies climate risks, enhancing toughness and avoiding losses from service interruptions. Resilience investments fortify populations, business continuity, and emergency operations essential for adapting to shifting climate realities.

Decarbonized Energy for Transportation

Electrification decentralizes vehicle charging and is amenable to integration with building infrastructure. Managed EV charging leverages off-peak HVAC capacity for grid services while vehicles stabilize building loads. Thermal management from batteries and electrical systems centralized at properties enhances efficiency and resilience, supporting transportation emissions reductions in tandem.

Renewable hydrocarbons, including hydrogen fuel, and different fleets eliminate tailpipe emissions when coupled with renewable production. Co-locating refueling and electricity generation at large buildings creates clean micro-mobility hubs. Shared autonomous vehicles reduce individual ownership, improving urban land use optimized around zero-emissions mass transit supportive of dense, walkable development patterns.

Digitalization Transforming Building Operations

Pervasive sensing, analytics, and automation optimize built environments. Continuous IoT commissioning detects inefficiencies from aggregated performance data, directing low-cost optimizations. AI assistants remotely monitor, diagnose, and resolve issues, preventing discomfort or waste. Integrated clean building certification benchmarks sustainability across portfolios.

Digital twin virtual replicas model retrofit scenarios assessing financial and emissions impacts prior investments. Blockchain substantiates emissions reductions monetized through carbon markets or sustainable branding. Advanced materials self-optimize, adjusting HVAC needs reduced. Big data applications transform management, lowering operating costs as sustainability rises fleet-wide.

Adopting Efficient, Eco-Friendly Technologies

Investing in the newest high-efficiency HVAC solutions qualifying for rebates pays dividends through utility incentives and long-term operating cost reductions. Pairing with renewable energy production maximizes savings further. For example, heat pump installations surpass code thanks to their efficiency, translating to lower monthly bills.

Demand response control integration enables additional earnings from curtailing loads during peak periods. Commissioning delivers persistent optimization, keeping systems performing as intended. Proper maintenance safeguards equipment lifespan, extending savings over decades.

Transitioning to lower global warming refrigerants in a phased approach mitigates risks from future bans while realizing upfront credits. Strategically timing projects to replace aging units with the greenest available alternatives maximizes future-proof emissions compliance.

Taking Advantage of Tax Incentives

Efficiency upgrades and on-site renewable energy generation furnish tax deductions, lowering total installation costs. The Residential Energy Efficient Property and Commercial Buildings Energy Efficiency tax deduction offers percentage-based credits for qualifying measures.

Depreciation schedules permit writing off investments over several years. Renewable energy property production tax credits provide direct savings per installed watt. Continuing education course deductions minimize outlays for mandatory license validations.

Thorough documentation substantiates tax benefit claims come review season. Consultation with financial advisors strategizes maximizing applicable incentives each year. Combining incentives multiplies savings potential from one undertaking.

Strict Documentation Practices

Organized recordkeeping establishes compliance standing should questions arise from inspections, litigation, or audits years later. Transaction forms, nameplate data, maintenance records, employee certifications, permits, and training certificates prove adherence to myriad regulations over time.

Digital record storage provides backup protection and accessibility from any location. Standard operating procedures institute consistent documentation practices across all company operations. Intake forms capture necessary service details, preemptively avoiding missing information complications.

Ongoing Regulatory Education

Keeping abreast of changing codes and best practices maintains quality work through refresher seminars and manufacturer-provided courses. However, leveraging qualifying training sessions as tax deductions offsets outlay. Employer-sponsored training minimizes individual costs.

Webinars provide convenient, low-cost learning wherever the internet exists. Professional association membership delivers timely regulatory alerts and discounts on continuing education credits, satisfying license obligations. Deductible conferences network with industry peers exchanging compliance expertise.

Rigorous self-assessments identify knowledge gaps, directing focus onto upcoming requirements. For example, refrigerant transition schedules necessitate familiarizing with new safety protocols well in advance through comprehensive certification programs. Proactive adaptation curtails risks from noncompliance issues.

Preventative Maintenance Programs

Well-maintained HVAC equipment operates at maximum rated efficiency and reliability, avoiding more expensive repairs or equipment failures down the line. Maintenance agreements spread costs predictably instead of unpredictable repair bills.

Regular tune-ups, filter replacement, coil cleaning, and component inspecting sustain optimal performance, justifying minimal expense. Contractor check-ins detect concerns early when inexpensive to address. Documentation substantiates diligent upkeep, potentially voiding warranty claims from neglected units.

Special Compliance Considerations

System alterations necessitating permits factor associated expenses judiciously against projected savings. For example, swapping a condensing unit may require conducting a Manual J load calculation and submitting construction plans. Weighing compliance obligations prudently sizes projects smartly.

Emerging technologies require vetting applicable regulatory approvals and pertinent exemptions. However, innovative, sustainable solutions satisfying multiple compliance targets deliver maximum long-term value for clients seeking industry leadership.

Certain industries impose specialized training, certification, or recordkeeping not typically affecting residential/light commercial settings. Understanding unique compliance nuances particular to relevant industries assists in avoiding violations.

Pursuing Sustainability Accreditation

Leadership in Energy and Environmental Design (LEED) certification, ENERGY STAR score benchmarking, and sustainable contractor credentials showcase regulatory outperformance, growing clientele seeking low-carbon partners. Compliance exceeding minimums ensures longevity through climate adaptation and resiliency.

Verification of sustainably sourced materials, zero emissions milestones, or distributed energy expertise furnishes competitive differentiators. Green building certifications may satisfy clients' internal sustainability goals or mandate familiar faces for future projects. Specialized credentials bolster opportunities within progressive industries prioritizing environmental stewardship.

Foresight proactively steering clear of compliance issues through optimized technologies and documentation practices preserves profitability. Strict adherence maintains licensing while tapping incentives offsetting investments. An abiding commitment to sustainability outstrips regulations, stimulating opportunities alongside minimized operating costs for decades.

CONCLUSION

Our journey through the comprehensive world of HVAC systems has come to an end, but hopefully yours is just getting started. After more than 20 years working in this industry, I'm more passionate than ever, and I hope this book has sparked your enthusiasm for refrigeration and climate control mastery!

We've covered an immense amount of ground together, from exploring HVAC fundamentals to outlining preventative maintenance practices, calculating efficiency upgrades, and so much more. By providing industry insights tailored for various experience levels, my aim was to illuminate the path to HVAC excellence, no matter where you're starting from.

I shared my personal experiences, beneficial shortcuts, handy diagnostic tricks, inspiring success stories, and crucial lessons learned to help you avoid common pitfalls. Whether strategizing upgrades as a homeowner or charting a career installation technician, the principles and skillsets outlined throughout The HVAC Bible are universally applicable across domains.

Most importantly, I hope this book has empowered you with in-demand abilities like:

- Accurately sizing HVAC equipment using precise calculations to optimize efficiency
- Comparing bids for installation/replacement with a discerning eye
- Independently inspecting, cleaning and maintaining all system components
- Mastering thermostat and smart controller programming for comfort/savings
- Pinpointing issues through strategic troubleshooting to minimize downtime
- Repairing minor problems yourself and discerning when professional help is needed
- Prioritizing indoor air quality alongside cost-effectiveness

- Grasping advanced techniques like geothermal and solar-powered HVAC
- Planning future-forward for regulatory changes and sustainability

With these skills and so much more now within your toolkit, you hold the keys to achieve substantial cost savings and perfectly calibrated comfort in any home or business through sound HVAC strategy.

I aimed to create the bible I wish existed when I first started out —alleviating all frustrations by covering everything modern climate control has to offer across residential and commercial applications. Consider this your HVAC Encyclopedia - a comprehensive reference guide to consult whenever questions arise in the future.

While technology will continue advancing, the foundational insights you now possess will serve you well. Having weathered recessions and fads, I'm confident the methodologies outlined equip you to navigate changes and innovations impacting HVAC for years to come through an educated lens.

My commitment is that this bible will pay dividends every time you leverage your expanded knowledge to make more informed HVAC decisions - saving money, time and hassle at every turn!

At the start of our HVAC journey, I emphasized the direct link between climate control mastery and greatly reduced anxiety about unpredictable repair costs or losing comfort. By learning key skills like maintenance and troubleshooting, you regain control over home or workplace environments no matter what disruptions arise. Consider it the ultimate form of assurance!

Of course, don't hesitate to refer back to relevant chapters whenever questions pop up. Post-it flag key sections or worksheets liable to prove handy down the line. Whatever your preferred style for putting new techniques into practice, this bible was written to be a well-worn resource as your capabilities expand!

While Part V concluded our walk-through indoors, I'd like to suggest one final bit of homework now that we part ways:

Go outside once it gets dark and lookup at the blanket of stars overhead. As you scan the skies, contemplate how the same fundamental principles regulating those atmospheric gases and combustible reactions are at work right underneath your roof.

Creating ideal conditions for human comfort relies upon many forces beyond our control. But after reading this book, far fewer climate-related uncertainties exist in your world. Now YOU dictate the forecast indoors thanks to being equipped with everything needed for HVAC mastery!

Of course, expanding competence inevitably surfaces more questions, so I welcome you to keep in touch with discoveries made putting recently gained wisdom into action. Perhaps connect with fellow readers to continue exchanging ideas and troubleshooting tricky scenarios together.

Most of all, promise me you'll pay forward anything you've learned here to help friends, family and neighbors reduce their home-related anxieties and expenses too. Assure them that maintaining comfortable conditions doesn't need to be complicated or costly armed with the right roadmap!

I may still spend nights gazing in wonder at constellations far off, but during the day, you'll find me doing what I love best — helping everyday people achieve climate control peace of mind. That liberating feeling never gets old!

Now spread your wings and fly towards HVAC mastery...I'll be cheering you on! Godspeed transforming spaces large and small through comfort and efficiency fine-tuning! Let the savings begin!

APPENDICES

GLOSSARY OF HVAC TERMS

Air Handler: The part of an HVAC system that circulates heated or cooled air throughout a home via a system of ductwork. It contains the blower fan and housing.

Boiler: A device used to heat water for use in a hydronic heating system. The heated water is circulated through a system of pipes and radiators or underfloor tubing to provide heat.

British Thermal Unit (BTU): A standard measurement of energy used in the heating and cooling industry. It quantifies the amount of energy required to heat or cool one pound of water by one degree Fahrenheit.

Coil: Made of copper or aluminum tubing, coils are used to absorb heat (in an air conditioning system) or expel heat (in a heating system). They often contain refrigerant that absorbs or expels heat.

Compressor: The pump in an HVAC system that pressurizes refrigerant gas as part of the refrigeration cycle. It is powered by an electric motor and often called the "heart" of the system.

Condenser: The outdoor portion of a central HVAC system that expels heat from the hot, pressurized refrigerant gas and condenses it to a cooler liquid state as part of the refrigeration cycle.

Condensate Drain Line: Removes and drains away condensed water vapor collected from the evaporator coil in air conditioning mode. This prevents excess moisture buildup.

Ductwork: Hollow metal tubes or channels used to direct conditioned air from the HVAC system throughout a home. The ducts connect the air supply to individual rooms.

Evaporator Coil: The indoor coil in an HVAC system responsible for absorbing heat from indoor air and causing the refrigerant inside the coil to evaporate. This has a cooling effect.

Expansion Valve: A component in the indoor section of an HVAC system that reduces the pressure of liquid refrigerant after leaving the condenser coil and regulates its flow into the evaporator coil.

Heat Exchanger: Transfers heat into or out of air directed into the HVAC ductwork by way of heated or cooled water, steam, or refrigerant. It keeps these substances separate from the airflow.

Heat Pump: An HVAC system capable of both heating and cooling by transferring heat into or out of an indoor space. One outdoor unit works with an indoor fan coil unit to provide heating, cooling, and usually air conditioning.

HVAC: Heating, ventilation, and air conditioning. Refers to technology and systems that achieve indoor environmental comfort and air quality.

Hydronic Heating: A system that heats a building by circulating heated water through a closed network of pipes. It often utilizes boilers and radiators or under floor heating.

Load Calculation: The HVAC industry term used to describe the manual process for determining the heating and cooling requirements for a home. This determines the capacity of equipment needed for the space.

Merv Rating: Measurement rating from 1 to 16 of an air filter's particle-capturing efficiency. Higher ratings capture more air particles and pollutants.

Psychrometrics: The engineering science concerned with studying the thermodynamic properties of moist air and uses these properties to analyze HVAC design and performance issues.

Refrigerant: The compound (e.g. R-22, R-410A) that circulates through an HVAC system to provide the heat absorption and heat rejection capabilities that make heating and cooling possible. They absorb and expel heat.

Register: A grille mounted over duct opening allowing conditioned air from the HVAC system to enter a room but preventing direct access inside the ductwork. Registers can control airflow direction.

Return Duct/Vent: An air duct system that directs air from within the home back to the furnace or air conditioner. This permits recirculation and filtering.

Split System: An HVAC system with separated components - typically an outdoor condenser and compressor coupled to a set of indoor furnace coils and air handling unit.

Thermostat: An HVAC control device that regulates a home's indoor air temperature by communicating orders to heating or cooling equipment to maintain a set temperature point. It may include a programmable schedule.

Tonnage: The amount of cooling capacity in tons provided by an HVAC unit. One ton equals 12,000 British Thermal Units (BTUs) of heat energy removed per hour. A typical home needs 1 to 5 tons of capacity.

Ventilation: The process of bringing fresh outdoor air into a space to replace stale or contaminated indoor air. This dilutes pollutants and controls indoor humidity.

Zone Control: Separate control of different "zones" in a home for HVAC management. Allows custom heating or cooling for select rooms or areas rather than the full house.

Zoning Damper: Motorized outlets within air supply ductwork that control airflow and temperatures in different areas (zones) of a home. They open and close automatically based on which zones require heating or cooling.

RECOMMENDED TOOLS AND EQUIPMENT

Having the right tools is essential for installing, maintaining, and repairing any HVAC system. Here are some of the most important tools and equipment to have on hand.

Essential Tools:
- Screwdrivers (standard, Phillips, torpedo)
- Pliers (needle nose, channel lock)
- Wire cutters & strippers
- Tube cutter
- Bolt cutters

- Hammer
- Utility knife
- Tape measure
- Level
- Flashlight
- Multimeter
- Clamp meter

Specialty Tools:
- Refrigerant gauges
- Airflow meter (anemometer)
- Infrared thermometer
- Carbon monoxide meter
- Psychrometer

Safety Equipment:
- Work gloves
- Safety glasses
- Ear protection
- Particle mask

For Maintenance:
- Condensate pump
- Wet/dry vacuum
- Coil brush
- Fin comb

Other Recommended Items:
- Ladder
- Tool belt
- Shop vac
- Cordless drill
- Extra filters
- Duct tape

HVAC Installation Checklist

Pre-Installation
○ Research local building codes
○ Take accurate room measurements
○ Determine necessary system capacity
○ Select HVAC equipment and contractor

Space Considerations
○ Ensure sufficient space for equipment and airflow
○ Allow room for future maintenance
○ Consider noise control techniques

Ductwork
○ Design efficient duct layout
○ Size ducts appropriately
○ Maintain minimum bend radius
○ Ensure proper support and joints

Electrical
○ Confirm appropriate power supply
○ Provide power source to unit
○ Install disconnect switch

Ventilation
○ Vent combustion gasses safely
○ Ensure adequate combustion air
○ Install intake and exhaust vents

Post-Installation
○ Pressure test for refrigerant leaks
○ Test ignite burner flame
○ Confirm proper airflow
○ Verify thermostat operation
○ Seal ductwork

Monthly HVAC Maintenance Checklist

○ Change air filters
○ Check and clean evaporator coil
○ Brush debris off condenser coils
○ Inspect electrical components
○ Lubricate blower motor if needed
○ Check thermostat operation
○ Verify proper temperature difference
○ Look and listen for unusual noises
○ Check condensate drain functionality
○ Inspect ductwork for leaks

HVAC Troubleshooting Checklist

No Cooling:
○ Check power supply
○ Inspect capacitors
○ Measure refrigerant charge
○ Verify condenser fan operation
○ Check for low airflow
○ Inspect relays and contactors
○ Test for refrigerant leaks
○ Look for frost pattern issues
○ Calibrate thermostat

No Heating:
○ Verify power supply
○ Inspect flame sensor
○ Check igniter or pilot light
○ Examine heat exchanger
○ Test heating elements
○ Inspect blower functionality
○ Measure airflow
○ Assess thermostat operation

Other Issues:

O Review error codes
O Check refrigerant lines
O Listen for odd noises
O Check condensate drain
O Inspect ducts for leaks
O Test for carbon monoxide

HVAC Energy Efficiency Checklist

Maintenance
O Regularly replace air filters
O Clear debris from outdoor unit
O Clean evaporator and condenser coils
O Inspect ductwork for leaks
O Verify refrigerant charge

Monitoring
O Track overall energy usage
O Log equipment run times
O Record start/stop frequency
O Monitor supply and return temperatures

Upgrades
O Install programmable thermostat
O Add zoning system
O Upgrade to higher SEER equipment
O Improve insulation and seals
O Consider smart home devices

Made in the USA
Middletown, DE
24 May 2024